HOLOGRAMS

HOLOGRAMS

A Cultural History

SEAN F. JOHNSTON
University of Glasgow

UNIVERSITY PRESS

Great Clarendon Street, Oxford, OX2 6DP,
United Kingdom

Oxford University Press is a department of the University of Oxford.
It furthers the University's objective of excellence in research, scholarship,
and education by publishing worldwide. Oxford is a registered trade mark of
Oxford University Press in the UK and in certain other countries

First Edition published in 2016
Reprinted 2016 (with corrections)

Published in the United States of America by Oxford University Press
198 Madison Avenue, New York, NY 10016, United States of America

British Library Cataloguing in Publication Data
Data available

Library of Congress Control Number: 2015944563

ISBN 978–0–19–871276–3

This book is dedicated to Libby, Dan, Sam and Ty.

PREFACE

Holograms reached popular consciousness during the 1960s and, over the following decades, evolved along with culture itself. No other visual experience is quite like interacting with holograms; no other cultural product melds the technological sublime with magic and optimism in quite the same way.

Why has its combination of attributes been perennially exhilarating and wondrous? Although the hologram was a product of its times, it was also wrapped in long-established cultural threads. Its impacts were conditioned by its antecedents and associations. This protective heritage helped make it impervious to technical disappointments, while ongoing innovation periodically revitalized its spectacular qualities.

This book traces the cultural roots of holograms to explore their influences and effects, assess their counteracting attractions and explain their persistence. It complements and extends my earlier book for Oxford University Press, *Holographic Visions: A History of New Science*. That volume provided an in-depth account of the creators and concepts behind the subject of holography as a scientific concept, an engineering tool and a business, and was based on extensive archival research and interviews with the originators of the field. Addressing historians of science and technology, holographers and sociologists of expertise, it recounted the trials of creating a new science, a new industry and a new profession from this remarkable concept.

The present book shifts the focus outwards. I turn from the *creators of holography* to the *consumers of holograms* in distinct cultural settings. This complementary emphasis enrols fresh audiences, too: readers attracted to visual and cultural studies, scholars concerned with media history, fine art and material studies and, most of all, cross-disciplinary audiences intrigued about how this ubiquitous but still-mysterious visual medium grew up in our midst and became entangled in our culture.

Sean Johnston
Dumfries, Scotland
June 2015

ACKNOWLEDGEMENTS

This book, like its companion, *Holographic Visions: A History of New Science* (Oxford: Oxford University Press, 2006), is the product of research supported by grants from the American Institute of Physics Center for the History of Physics, the British Academy, the Carnegie Trust for the Universities of Scotland, the Royal Society, Shearwater Foundation and the University of Glasgow. I am grateful to archival staff at the Bentley Library, University of Michigan, Ann Arbor MI; the MIT Museum, Cambridge MA; National Museums Scotland, Edinburgh; the Niels Bohr Library and Archives, College Park MD; the Science Museum Library, London; and, the Smithsonian Institution, Washington, DC. I also thank the museum curators, collectors, entrepreneurs, gallery managers, holographic scientists, engineers, artists, administrators and enthusiasts who have shared with me their time, experiences and holograms.

Particular assistance with information is gratefully acknowledged from Iñaki Beguiristain, Hans Bjelkhagen, Patrick Boyd, Jan Broeders, Tove Dalenius, Frank DeFreitas, Art Funkhouser, Nathan Gonzales, Darran Green, Skip Henderson, Samuel B. Johnston, Gretchen Morris, Craig Newswanger, Martin Richardson, Jonathan Ross, Itsuo Sakane, Juris Upatnieks, Anita West and Gary Zellerbach.

Finally, I thank Sönke Adlung (Senior Commissioning Editor, Physics) and his colleagues at Oxford University Press, who have carefully guided the manuscript to publication.

CONTENTS

1
Introduction

Is it possible to write the biography of an object? Like all of us, the hologram was born full of potential and has been shaped over its lifetime by its cultural environments. A biography can reveal those innate traits and how they blossomed in different settings. This book argues that in our contemporary technologized world, artefacts and people intermingle so pervasively that any history must consider both.[1]

Who is this central character, then? The hologram is an unusual invention: an innocuous optical device that carries within it the ability to generate three-dimensional imagery. But since its conception after the Second World War, it has assumed a bewildering variety of forms and purposes. The hologram has been a crucial element in military hardware, engineering procedures, works of art and video games. For a time, it promised to be the successor to the photograph. And its varying visual impact has been paralleled by a fluctuating symbolic value, alternately representing scientific progress, technological illusion and personal security.

How do we capture the identity of the hologram and an understanding of how it became embedded in its cultural niches? Before beginning, we have to recognize that this is not a conventional biography. The hologram does not fit the traditional life story—or scientific narrative—of cossetted childhood, growing maturity and recognized success. It is more akin to a personality developed in a series of foster homes in exotic locales, and a subsequent life juggling jobs and relationships. If the hologram were a person, it might be characterized as a case of multiple personalities, flitting from one unsettled identity to another. Putting a recognizable face on it requires us to follow the hologram through its different environments, to see how it has been accepted and employed.[2]

[1] The sociologist Bruno Latour argues that 'things' must be considered as social actors just as people are. Artefacts are components of networks also made up of people, institutions and procedures, and play an important role in the evolution of socio-technical beliefs, choices, systems and activities. In a different sense, Lorraine Daston and colleagues have argued that scientific concepts, too, can be tracked as philosophical 'objects' that are born, prosper and die as a result of such forces. See, for example, Latour, B., *Science in Action: How to Follow Scientists and Engineers Through Society* (Milton Keynes: Open University Press, 1987); Daston, L. (ed.), *Biographies of Scientific Objects* (Chicago: University of Chicago Press, 2000).

[2] The psychological analogy of multiple personalities has some kinship to analytical understandings of relevant social groups in a model of sociology of science and technology known as 'social construction of technology', developed by Wiebe Bijker and Trevor Pinch, and also to the concept of 'research technology' developed by Terry Shinn and others. See, for example, Pinch, T. J. and W. E. Bijker, 'The social construction of facts and artifacts: or how the sociology of science and the sociology of technology might benefit each

Holograms have inhabited diverse environments. Their predecessors (for holograms had a long family history) entertained and educated the public with spectacular imagery. But the first holograms were conceived as components of sophisticated systems, and occupied narrow audiences. They later surfaced as a scientific marvel in their own right, and later still as an expression of the counterculture and of consumer magic, only to disappear again into the murky world of counterfeiting.

The form, function and meaning of this elusive medium have been recast repeatedly. Holograms channelled scientific aspirations, institutional needs, individual expression and popular understandings. Nurtured by distinct subcultures in diverse environments, they were envisaged during their first two decades as a breakthrough for seeing the ultra-small, or alternatively as a secret and sophisticated alternative to the primitive military computers of the day. During the 1960s the hologram made its public debut for engineers as an awe-inspiring three-dimensional imaging medium and exquisite analytical tool. For diverse audiences, this magical creation represented the essence of modernity and the future itself. In new cultural environments it mutated into an art form, a dazzling medium for advertising and education and a foil for counterfeiters. For a while it was domesticated to hang on living-room walls. And perhaps most compellingly, the hologram spread through virtual space: it captured popular imagination as an imminent entertainment medium and became a mainstay of science fiction alongside wormholes and warp drives. For the following generation, holograms were associated with videogames and entertainment spectacles. Through holograms new cultural groups, values and expressions co-evolved and coalesced. New viewers, in fresh settings, repeatedly made holograms their own and invested them with new meanings. Engineers, artists, hippies and hobbyists have played with, and dreamed about, holograms.

The attributes of the holographic image are a sensitive indicator of those cultural environments. This book examines the technical attractions and cultural uses of the hologram, how they were conditioned by what came before them, and how they have matured over a half-century to shape our notional futures. Today, holograms are in our pockets (as identity documents) and in our minds (as gaming fantasies and 'faux hologram' performers). Why aren't they more often in front of our eyes?

1.1 CULTURE AND TECHNOLOGY

The role of culture underpins artistic and creative fields, but is often overlooked in scientific and technical subjects. Culture incorporates dimensions that include collective

---

other', in: W. E. Bijker, T. P. Hughes and T. J. Pinch (eds.), *The Social Construction of Technological Systems* (Cambridge, MA: MIT Press, 1987), pp 17–50; Bijker, W. E., T. P. Hughes and T. J. Pinch, *The Social Construction of Technological Systems: New Directions in the Sociology and History of Technology* (Cambridge, MA: MIT Press, 1990); Shinn, T. and B. Joerges, 'The transverse science and technology culture: dynamics and roles of research-technology', *Social Science Information* 41 (2002): 207–51 and Joerges, B. and T. Shinn (eds.), *Instrumentation: Between Science, State and Industry* (Dordrecht: Kluwer Academic, 2001).

belief, community values, popular fashion, customs, philosophical or religious convictions, shared aesthetic sensibilities and forms of creative expression such as art and music. Holograms, and three-dimensional imagery more generally, have influenced all of these domains, and cultural studies provide comprehensive perspectives that reveal how these elements interact. Indeed, holograms can be appreciated as a visual medium that, for a time, satisfied the enduring cultural appeal of three-dimensionality—alongside classical sculpture, Victorian optical toys, 3-D TV and 3-D printers.

The relevance of cultural concepts may not always be obvious, but they can enrich our understandings of how we engage with technologies. The media theorist Marshall McLuhan argued that new media technologies can masquerade as, or borrow from, their predecessors.[3] A cultural history of holograms must consequently trace these antecedents and reveal their influences.

Any one of these perspectives—scientific, technological, artistic and economic—may restrict vision, too, in just the way that the Victorian stereoscope provides a limited sense of reality, unable to view the parts of the scene masked by others. Disciplinary experts, as much as members of the wider public, may invest technologies with their expectations and prior associations. Holographer Stephen Benton, for example, expressed a widespread belief when he suggested that new technologies evolve inexorably: 'A satisfying and effective three-dimensional image is not a technological speculation, it is a historical inevitability'.[4]

But historians of technology can find many counter-examples. Indeed, three-dimensional imaging technologies as a class have had mixed fortunes in generating satisfaction. Some, like nineteenth-century stereoscopes, influenced popular culture for decades, but others have been dismissed by consumers with scarcely a second glance. And even stereoscopes declined as a mass medium for Edwardian audiences. 3-D cinema has mushroomed periodically in popularity, only to decay again within a decade or two. And, with the growing union between electronics, computing and optics since the 1980s, a variety of sophisticated imaging technologies have enthused their engineering advocates but sometimes made little impact beyond their specialist communities. Virtual reality headsets, for example, were a flash in the pan during the 1990s. Recognizing the historical inevitability of holograms or even three-dimensional imaging, then, may be more a matter of collective belief than a reliable anticipation.

Explaining popular engagement with technologies requires more than technical analysis. Some imaging technologies, of course, have clear technical limitations. Imaging systems that do not provide all the attributes of three-dimensional perception (stereopsis)—or, even worse, force them to compete within our visual system—are likely to be found wanting in some circumstances. But the rise and fall of such technologies also reveals the shifting interplay between perception, collective desire and personal

[3] McLuhan, M., *Understanding Media: The Extensions of Man* (New York: McGraw-Hill, 1964).

[4] St.-Hilaire, P., M. Lucente, J. D. Sutter, R. Pappu, C. D. Sparrell and S. A. Benton, 'Scaling up the MIT holographic video system', *Proceedings of SPIE* 2333 (1995): 374–80.

experience. For a time, those technologies were appealing, seductive and exciting—only to be rejected later when they no longer mapped onto new criteria drawn from prevailing fashions and cultural judgements. Such technologies did not, or did not for long, resonate with wider community values.

1.2 WHY CULTURAL HISTORY?

The cultural studies approach complements and enhances technological evaluations. Where possible, it measures what can be quantified—e.g. the number of stereoscopes sold or the attendance figures for hologram exhibitions—and bases hypotheses at least partly on such data. But it also adopts other approaches to capture popular attitudes and characterize community views. As the name suggests, it is informed by history, anthropology and other forms of cultural study to investigate the shared social, political and intellectual environments of groups of people.

The term 'cultural history' had its origins in the 1920s and became a particularly popular historical perspective from the 1990s.[5] Its viewpoint can be unsettling. It does not focus primarily on events and key figures or even technological 'breakthroughs', which are more the subject matter of past generations of military, political or scientific histories. Nor does it concentrate on social groups and systemic causes (the domain of social history), or even on the rational refinement of intellectual ideas (the remit of intellectual history). Instead, cultural history directs attention to the meanings and motivations given by individuals or collective forces that make up a culture or subculture. Often focusing on popular culture—mass media, fashion, new technologies and evolving customs—it can probe deep societal structures via tools such as critical theory, or alternatively give a perceptive but unfamiliar characterization of changing ways of life. This discipline-crossing approach can be both fertile and disorienting, encouraging the re-examination of our shared perceptions and beliefs.

But some have argued that with the rising popularity of cultural history has come a shift away from analytical precision to mere entertainment for broader audiences.[6] In its most superficial form, cultural history may merely narrate a tale of earlier vogues without explanation of how or why they came and went. Such 'virtual witnessing' is liable to be inadequate and misleading, providing nostalgia but few insights. In such cases, the potential for better understanding cultural forces is lost.

A similar point has been made concerning trends in studies of science, technology and society (STS), a field that originally identified itself in opposition to established

[5] An early illustration of the varied approaches and meanings is given by Ware, C. F. (ed.), *The Cultural Approach to History* (New York: Columbia University Press, 1940). Good recent introductions to the subject include Arcangeli, A., *Cultural History: A Concise Introduction* (New York: Routledge, 2012) and Green, A., *Cultural History* (Basingstoke: Palgrave Macmillan, 2008). For an evolving and increasingly powerful database that tracks word and phrase popularity, see Google ngrams, http://books.google.com/ngrams.

[6] Pihlainen, K., 'Cultural history and the entertainment age', *Cultural History* 1 (2012): 168–79.

understandings of the modern world, but which increasingly has been appropriated by policy-makers, business analysts and even corporations.[7] This account does not focus on business history: attributions of holography's significance based on economic activity, numbers of practitioners and technical progress have fuelled decades of conferences, but may reveal relatively little about the depth of cultural engagement. Nor does it rehearse the development of the intellectual discipline (holography) and its profession (holographers), the subject of my previous work.[8] Instead, the book complements such perspectives. I argue that cultural history has the capacity to illustrate how shared representations come to be, and to explain the implications of this process for wider culture.

Cultural history and STS are interdisciplinary, combining insights from sociology, philosophy, literary studies and ethnology. Both of these research perspectives focus their attention on previously under-studied human activities and products, and on alternative framings by their historical participants. Over the past generation, STS has explored how social groups use (or do not use) particular technologies, and how communities and technologies may co-evolve together.[9] The ubiquity of social construction is an underlying refrain. Both fields have employed micro-historical studies to explain how local circumstances framed broader understandings. And they complement each other. Where cultural history has vaunted popular culture as an important domain of study, science and technology studies have focused on the complementary practices of scientists, engineers, technologists and organizations responsible for envisaging, designing and sustaining the trappings of modernity that are at the heart of contemporary popular culture itself.

There is an anthropological dimension to this study, too. Over a dozen years, I have interacted closely with hologram producers, hologram enthusiasts, promoters, collectors and consumers. Their domains included studios, workshops, homes, labs, factories, offices, galleries, conferences and classrooms. These conversations, interviews and surveys supplied narrative descriptions that revealed users' engagement with holograms. This experience 'in the field' complemented more distant and totalizing historical and technical perspectives. At the heart of these methods is the notion of 'thick description' expounded by anthropologist Clifford Geertz. Geertz sought to explore the full context surrounding beliefs and behaviours, an approach adopted increasingly by social scientists.[10]

[7] Woolfgar, S., C. Coopmans and D. Neyland, 'Does STS mean business?', *Organization* 16 (2009): 5–30. STS is also an acronym for the closely-related cluster of scholars of Science and Technology Studies.

[8] Johnston, S. F., *Holographic Visions: A History of New Science* (Oxford: Oxford University Press, 2006).

[9] Social Construction of Technology (SCOT), developed from the early 1980s, is a prototype for this form of analysis. Subsequent focus on co-production of user groups and technologies offers a refinement of the original concept. See, for example, Oudshoorn, N. and T. Pinch (eds.), *How Users Matter; The Co-Construction of Users and Technology* (Cambridge, MA: MIT Press, 2005).

[10] Geertz, C., *The Interpretation of Cultures: Selected Essays* (New York: Basic Books, 1973). An analogue of this approach in technology studies is investigation of the 'seamless web' of economic, political, technological and social interests making up a system; see, for example, Hughes, T. P., 'The seamless web: technology, science, etcetera, etcetera', *Social Studies of Science* 16 (1986): 281–92.

The chapters focus on qualities that have served anthropologists as cultural markers, such as magic, work, consumption and celebration. Holograms evoked and satisfied long-standing cultural desires. But the book also explores those more familiar to other disciplines—such as notions of productivity, innovation and secrecy—each of which gained importance during the twentieth century.

Anthropologists have often focused on *things*, too—products and precious objects both functional and symbolic—as a route to understanding their human users. Indeed, nineteenth-century anthropologists often 'objectified' knowledge about other cultures via collections of their tools, garments, art and items of ritual. This was only gradually supplemented by more informative approaches.[11] As Ruth Schwartz Cowan suggested, however, an effective starting point for modern historians and sociologists of technology is to follow the consumption of goods. This book heeds that approach with caveats: holograms have always represented a sparsely occupied 'consumption junction'.[12] My explorations of material culture are wary of relying on popularity alone as a justification for my coverage.

A history of holograms also demands attention to more ephemeral cultural indicators. Technologies may be invested with concepts that influence our perception of their physical characteristics. Shared memories can constrain how we interpret new inventions; shared dreams can shape how we exploit new opportunities.[13] I explore the role of popular literature in creating environments for creativity and innovation. And the imagery embodied in holograms reveals their roots in wider visual culture through the twentieth century.[14]

Holograms are sensitive cultural products that have inspired and shaped diverse subcultures. They have inhabited every niche of modern life: from mundane objects of kitsch fashion and personal identification, to visual media that express our exhilaration with the present, to fictional technologies that encapsulate our longings for a better world.

[11] Buchli, V., 'Introduction', in: V. Buchli (ed.), *The Material Culture Reader* (Oxford: Oxford University Press, 2002), pp 1–22.

[12] Cowan, R. S., 'The consumption junction: a proposal for research strategies in the sociology of technology', in: W. E. Bijker, T. P. Hughes and T. J. Pinch (eds.), *The Social Construction of Technological Systems* (Cambridge, MA: MIT Press, 1987), pp 261–80.

[13] On cultural memories, see Connerton, P., *How Societies Remember* (Cambridge: Cambridge University Press, 1989). On imagined futures, see Corn, J. (ed.), *Imagining Tomorrow: History, Technology and the American Future* (Cambridge MA: MIT Press, 1986).

[14] For useful overviews, see Sturken, M. and L. Cartwright, *Practices of Looking: An Introduction to Visual Culture* (Oxford: Oxford University Press, 2009); Walker, J. A. and S. Chaplin, *Visual Culture: An Introduction* (Manchester: Manchester University Press, 1997) and, more critically, Elkins, J., *Visual Studies: A Skeptical Introduction* (New York: Routledge, 2003). As Elkins observes, the imagery of science is a topic relatively little studied. Important contributions to understanding the visual cultures of scientists, engineers, educators and printing experts have been explored, however, in Hentschel, K., *Visual Cultures in Science and Technology: A Comparative Perspective* (Oxford: Oxford University Press, 2014); Hentschel, K., *Mapping the Spectrum: Techniques of Visual Representation in Research and Teaching* (Oxford: Oxford University Press, 2002); Hentschel, K. and A. D. Wittmann (eds.), *The Role of Visual Representations in Astronomy: History and Research Practice* (Frankfurt am Main: Verlag Harri Deutsch, 2000); and Jones, C. A. and P. Galison (eds.), *Picturing Science, Producing Art* (New York: Routledge, 1998).

Holograms reveal deep-rooted notions of modernity that have not just reflected contemporary understandings, but shaped our imagined futures, too.

1.3 AUDIENCES AND ASSESSMENTS

This book applies the integrative vision of cultural studies to the distinct consumers of imaging technologies—professional, amateur and popular—to explain how holograms found a place. To do so, it explores cultural domains that developed over more than a century.

Part A (Chapters 2 and 3) examines the aspects of visual culture that later supported a role for holograms. While their principles and technical innovations were a revelation during the 1960s, the background to the appeal of holograms had a long prehistory, with cultural associations that developed from the mid-nineteenth century. Earlier phases introduced themes that recur throughout this account. Among them was the interplay of professional and enthusiast cultures, and of enthusiast and popular cultures. Where professionals explored the possibilities of novel imaging technologies, enthusiasts appropriated them for new purposes and popular audiences consumed them. Professional cultures jostled and merged with innovative amateurs, and popular audiences extended their visual grammar. The nature and goals of imagery—and eventually holograms—co-evolved with their audiences. Evolving notions of imaging science, visual entertainment and visual education played an important role.

Those two introductory chapters focus on three dimensions of visual culture that were crucial in fostering an environment for holograms. First, imagery was reconceived as a *scientific* product; second, it was *democratized*; and third, its potential for *visual surprise* and entertainment was nurtured. The book introduces the professional cultures of engineers, technicians, artists and publishers, but it is increasingly dominated by successive waves of popular engagement.

Part B explores the origins of holograms and how professionals, enthusiasts and wider audiences made sense of them. Chapter 4 focuses on the increasingly covert engineering cultures after the Second World War that conceived and nurtured the first holograms, while Chapters 5 and 6 explore the remarkably universal attractions of holograms that relied on appeals of the past and of the future, respectively. Ancient magic and anticipations of progress converged to evoke remarkably persistent visions and rhetoric.

In Part C, five chapters examine how holograms were absorbed in specific cultural niches. Chapters 7–9 trace the appropriation of holograms by technical enthusiasts, technical countercultures, and artists, critics and collectors of aesthetically appealing holograms.

The final two chapters focus on the period in which popular uses and understandings of holograms were shaped for contemporary audiences, namely the early 1980s and beyond. In some respects, it is a tale of cultural decline, with waning appeal for holograms in both art and engineering. In technical and economic terms (Chapter 10) and in popular imagination (Chapter 11), however, holograms reached mass audiences for the first time.

To return briefly to the metaphor of biography, this book explores cultural environments in which the hologram developed, found a home and made an impact. It illuminates those contexts by focusing on how the hologram, in alternate guises, was employed and valued, how its adopters came to understand it and relate it to their world and how it was invested with functional and symbolic qualities that were distinctive to each of its environments.

But such a biographical approach may enlighten only those contexts in which the star shone. Of potentially equal interest is a complementary narrative structure, one that recounts the cultural contexts in which holograms did *not* prosper or have enduring influence. These temporary footholds, short dalliances and abortive initiatives were common in the history of holograms and reveal much about the incommodious environments and communities where they ventured unsuccessfully.

Between the lines, then, this is a story of cultural relativity. As experienced with earlier visual media, not everyone fell in love with holograms. Modern culture is a combination of both the popular and the unpopular. Some ideas and expressions prove fertile and fashionable; others wither or become stranded in no man's land. And where favoured concepts explode into ubiquity, others become unutterable. Holograms literally and figuratively straddle the light and shadow, representing pop culture and what could be dubbed 'unpop' culture.[15] This mixed story of advance and retreat, openness and concealment, progress and failure is not yet a prevalent style of recounting either biography, technological development or cultural history, but arguably provides a balance that more recognizably reflects our personal and collective experiences.[16]

The aim of this book, then, is to explore a combination of cultural history and history of technology that paints the nuances of influence, impact and malleability. During the late twentieth century, holograms and popular culture developed together and shaped each other. Understanding the wider implications of this co-evolution is essential for riding the succeeding waves of our technological culture.

[15] While the phrase 'pop culture' grew in popularity from the early 1960s, the counter-phrase 'unpop culture' has hitherto been restricted mainly to describing contemporary niche and special-interest performance arts.

[16] Much of my research has focused on historical studies of fields that inhabit the hinterland between success and failure in techno-scientific culture, an interstitial zone that I dubbed 'peripheral sciences'. See Johnston, S. F., 'In search of space: Fourier spectroscopy 1950–1970', in: B. Joerges and T. Shinn (eds.), *Instrumentation: Between Science, State and Industry* (Dordrecht: Kluwer Academic, 2001), pp 121–41; Johnston, S. F., *A History of Light and Colour Measurement: Science in the Shadows* (Bristol: Institute of Physics Publishing, 2001); Johnston, S. F., *The Neutron's Children: Nuclear Engineers and the Shaping of Identity* (Oxford: Oxford University Press, 2012).

PART A

Visual Culture and Modernity: The Backstory to Holograms

Holograms were an invention of the modern era, the last century-and-a-half of human activity when we fell hopelessly in love with inventions and the powers and pleasures they offered.[1] By the mid-twentieth century, when the first holograms were created, the cultural appeal of modernity was near its zenith. Over the following decades, holograms became a short-hand for illustrating the modernist worldview and its technological future.

The key features of this attraction were becoming apparent decades before holograms were even invented: visual imagery became increasingly scientific, spectacular and accessible.

[1] While having subtly different meanings and expressions in different disciplines, *modernism* is used here to label the philosophical movement which broadly rejects the values of past culture and seeks to reinvent all forms of human product and practices—including literature, social organization, architecture and art—along rational lines. Its central theme is confidence in the progressive potential of scientific methods and technologies.

2

Scientific Imagery and Visual Novelty

Nineteenth-century visual culture united high science with popular spectacle—a potent combination.[1] Inventors and inveterate innovators between the 1830s and 1870s developed a bewildering variety of photographic and printing processes for creating images. Over that period and beyond, the practical arts of imaging became increasingly allied with scientific understandings to yield spectacular visual experiences. As industrial creativity became culturally visible, innovation itself became a theme of new imaging industries.

2.1 MODERN ALCHEMY

The earliest of these imaging techniques were slow and meticulous to carry out, and often required the use of dangerous chemicals. *Heliography*, invented by Joseph Nicéphore Niépce in the early 1820s, employed a natural form of bitumen as a photosensitive material coating a pewter sheet and demanded several hours of exposure to a sunlit scene focused by a lens. After exposure, the sheet was washed with lavender oil, hardening the exposed areas and removing the unexposed areas of the bitumen to reveal a visible image. By contrast, the *daguerreotype* (1829), the earliest practical method devised by Niépce and collaborator Louis Daguerre, required the polishing a silver-coated copper plate, subjecting it to iodine vapour and then exposing it for several minutes to the light focused by a lens to record a latent image. Finally, the exposed plate would be subjected to mercury fumes to 'develop' a visible image. Each step required careful mechanical preparations and chemical practice. Later use of bromine and chlorine fumes for sensitizing the plates made the process even more dangerous for inexperienced photographers.

Subsequent methods relied on dissimilar chemical processes and yielded distinctive image characteristics. The *calotype*, announced by William Henry Fox Talbot in 1841,

[1] Precursors for this were the nineteenth-century fashion for scientific entertainments such as public lectures on chemistry and electricity. See, for example, Morus, I. R., *Frankenstein's Children: Electricity, Exhibition, and Experiment in Early-Nineteenth-Century* (Princeton, NJ: Princeton University Press, 1998); Fara, P., *An Entertainment for Angels: Electricity in the Enlightenment* (Cambridge: Cambridge University Press, 2002). These sought to educate, entertain and, at times, surprise their audiences, features that later became associated with photographs and holograms.

was created by painting a solution of silver nitrate onto writing paper, drying and dipping into a bath of sodium iodide before drying again. Just prior to exposure in the camera, the paper was brushed with a combination of silver nitrate, gallic acid and acetic acid, and after exposure brushed again with the same mixture while heating it gently to reveal the latent image. Finally, the image was made stable, or 'fixed', by immersion in hot sodium thiosulfate and then dried. This process yielded a negative image on paper. In order to create the desired positive image, a second piece of prepared paper was placed underneath the 'negative', exposed to sunlight and then processed as before. The two-step method allowed for multiple copies, but the paper imperfections made the image quality poorer than daguerreotypes. By contrast, the *tintype*, described by Adolphe-Alexandre Martin in 1853, employed a coating of silver salts in gelatine or collodion (a nitrocellulose compound) on a darkened thin plate of iron. When chemically developed in the same way as calotypes, tintypes produced a positive image because they were viewed by reflection against the dark background of the plate. The same scheme was used for *ambrotypes*, a similar process to generate an image on glass that would be viewed against a black background.[2]

A social dimension co-developed with the technology. Photographic practice was shaped by arcane chemistry and bulky cameras. Photographers became adept at preparing plates and performing their chemical treatments on site, a demanding skill that restricted the practice and consumption of photography. The commercial expansion and popular uptake of photography was aided by the development of the dry-plate process during the 1870s. Chemical treatments for gelatine made the plates robust and, better still, available from dedicated companies. These varied photographic technologies, then, co-evolved with specialist expertise, constraints on viewing and reproduction and opportunities for wider public engagement.[3]

Scientific culture was reshaped by photography, too. This growing collection of practical arts encouraged scientific analysis to explain the individual mechanisms involved and to improve and generalize the processes. The commercial potential made this deeper knowledge appealing to firms seeking to exploit new patents and new manufacturing markets. Empirical results founded on lucky accidents or methodical exploration could be the making of a successful firm, but theoretical understandings assured steady progress and a longer-lived competitive edge.[4]

[2] For a contemporary example of the art, see Burbank, W. H., *The Photographic Negative—written as a practical guide to the preparation of sensitive surfaces by the calotype, albumen, collodion, and gelatin processes, on glass and paper, with supplementary chapters on development, etc* (New York: Scoville Manufacturing Co, 1888). For general histories of photography, see Gernsheim, H. and A. Gernsheim, *The History of Photography* (Oxford: Oxford University Press, 1955); Lloyd, V., *Photography: The First Eighty Years* (London: Colnaghi, 1976); and Marien, M. W., *History of Photography: A Cultural History* (London: Laurence King, 2002).

[3] Batchen, G., *Burning with Desire: The Conception of Photography* (Cambridge, MA: MIT Press, 1999).

[4] Such steps in experimental photochemistry were repeated a century later by hologram enthusiasts and entrepreneurs. During the 1970s, after a century of stability, the chemistry of light-sensitive emulsions was explored again by photographic firms and enthusiasts to improve the quality of holograms, and led to a similar expansion of markets and audiences.

2.2 SCIENTIFIC IMAGING FOR ENTHUSIASTS AND PROFESSIONALS

Continued development of photography (or chemical imaging, to situate the field more generically) relied on enthusiasts' zeal as much as on business acumen. This will be a recurring theme in subsequent chapters: technical development, while dominated episodically by firms and institutions, proved to be driven just as reliably by a kind of enlightened amateurism.[5]

By their nature, such enthusiasms tend to be individualistic.[6] Two cases of early imaging science are exemplary: first, the indefatigable William Abney and, second, Ferdinand Hurter and Vero Charles Driffield. Both provide recognizable templates for later hologram enthusiasts and professionals.

William de Wiveleslie Abney (1843–1920) was a typical, if uncharacteristically energetic, Victorian technical enthusiast. In an environment that lacked a system of academic posts and any government commitment to funding scientific education and applied research, he was the very model of a scientific amateur.

Abney's career began with the Royal Engineers and in successive posts as a chemical assistant and instructor of telegraphy, he became responsible for a school of chemistry and photography at Chatham School of Military Engineering and Inspectorate of School Science at the Science and Art Department located at South Kensington, London.[7] His working life was devoted to visual science, photographic research and educational publications.

An energetic networker, Abney gave courses of public lectures on photography and the scientific measurement of light and colour, both of which led to popular books. Editor of *The Photographic Journal* (London) from 1876 until his death, he was also a prolific contributor to numerous photographic, astronomical and scientific publications. In 1881 he introduced the first sensitive photographic emulsion based on gelatine—an innovation rapidly pursued by manufacturers. Abney steered scientific and technical societies, too: he was elected president of the Royal Photographic Society four times between 1892 and 1905, president of The Astronomical Society from 1893 to 1895 and of the Physical

[5] On urban examples of popularized science during the period, see Sheets-Pyenson, S., 'Popular science periodicals in Paris and London: The emergence of a low scientific culture, 1820–1875', *Annals of Science* 42 (1985): 549–72.

[6] Recent work on contemporary and historical enthusiasms is relevant here. See, for example, Geoghegan, H., 'Emotional geographies of enthusiasm: belonging to the Telecommunications Heritage Group', *Area* 45 (2012): 40–6; Morris, C. and G. Endfield, 'Exploring contemporary amateur meteorology through an historical lens', *Weather* 67 (2012): 4–8; Jankovic, V., *Reading the Skies. A Cultural History of the English Weather, 1650–1820* (Manchester: Manchester University Press, 2001). The sociological dimensions of amateurism have been explored by Robert Stebbins in, e.g., Stebbins, R. A., 'Science amateurs? Rewards and costs in amateur astronomy and archaeology', *Journal of Leisure Research* 13 (1981): 289–304; Stebbins, R. A., *Amateurs, Professionals and Serious Leisure* (Montreal: McGill University Press, 1992).

[7] On the nineteenth-century environment that spawned careers from technical enthusiasms, see Cardwell, D. S. L., *The Organization of Science in England* (London: 1972) pp 179–84 and Meadows, J., *The Victorian Scientist: The Growth of a Profession* (London: British Library, 2004).

Fig. 2.1 Royal Photographic Society judging panel, 1902, with William Abney on far right (RPS collection, National Media Museum, UK, Science & Society Picture Library).

Society between 1895 and 1897 (Figure 2.1). His cross-fertilization of astronomy, physiology, photography and physics introduced many of his scientific contemporaries to photography as a scientific tool and a subject of science in its own right.

Abney's missionary zeal to promote the science of imaging was not easy to popularize. He lamented that 'of 25,000 people who took photographs not more than one cared for, or knew anything about, the why and wherefore'.[8] Even his peers demurred: his presidency of the London Photographic Society in the 1890s prompted one chronicler to note that 'the meetings became still duller, and *The Photographic Journal* was devoted almost exclusively to scientific aspects of photography'.[9] The scientific investigation of photography, despite Abney's proselytizing, remained a narrow activity into the twentieth century.

But scientifically designed products offered business opportunities, and the professional activities of Ferdinand Hurter (1844–1898) and Vero Charles Driffield (1848–1915) became typical of commercial research. Hurter was a Swiss industrial chemist who made his career at an alkali plant in northern England, and Driffield was an engineer there.

[8] 'Obituary notice: William de Wiveleslie Abney', *Proceedings of the Royal Society* A99 (1921): i–v; Butterworth, H., *The Science and Art Department, 1853–1900*, PhD dissertation, Sheffield (1968).

[9] Gernsheim, H. and A. Gernsheim, *The History of Photography* (Oxford: Oxford University Press, 1955), p.256.

From the late 1870s they published a series of papers on sensitometry (the sensitivity of photographic materials to light) and densitometry (the consequent darkening of the developed material). Their analyses allowed photographic properties to be characterized, reliably manufactured and used. Such work transformed photography from a black art into a scientific field.[10]

Photographic lenses were similarly subjected to systematic research. The mathematics of lenses and mirrors had been studied since the Middle Ages; by the mid-nineteenth century, lens design and production of optical glasses were well understood and, in the face of growing competition, optical manufacturers innovated by mathematically calculating and systematically testing improved lenses.[11] Amateur photographers, a growing contingent, became versed in the advantages of particular lens designs for yielding more faithful photographic images and shorter exposure times.[12]

The commercial development of lenses yielded remarkable generalizations of scientific knowledge, too. Perhaps the least appreciated but most significant of them was the theory of optical imaging developed by Ernst Karl Abbe (1840–1905), a physicist who had worked at the Göttingen Observatory and University of Jena, and later became a professor there while employed as Research Director at the Zeiss Optical Works. Abbe systematically studied optical glasses, lens design and imaging. He derived mathematical relationships to dramatically improve the quality of microscopes and to determine the ultimate limitations of resolution. His findings later had a direct bearing on the theory of holograms.[13]

Scientific research in industry exploded at the turn of the twentieth century. The merger of small companies and increasing international competition encouraged the formation of science-based industry founded on industrial laboratories. Optical, chemical and electrical companies were among the first industries to employ professional scientists. Abbe

[10] See, for example, Hurter, F. and V. C. Driffield, 'Photochemical investigations and a new method of determination of the sensitiveness of photographic plates', *Journal of the Society of Chemical Industry* 9 (1890): 455–69. A graphical tool, dubbed the Hurter & Driffield or H&D curve, became the standard technique for characterizing photographic materials over the next century. Regarding the limited attention given to scientific investigation within the photographic industry, see Edgerton, D. E. H., 'Industrial research in the British photographic industry, 1879–1939', in: J. Liebenau (ed.), *The Challenge of New Technology* (Aldershot: 1988), pp 106–34.

[11] Rashed, R., 'A pioneer in anaclastics: Ibn Sahl on burning mirrors and lenses', *Isis* 81 (1990): 464–91; Sabra, A. I., *The Optics of Ibn al-Haytham (Books I-V)* (Kuwait: National Council for Culture, Arts and Letters, 1983–2002); Newton, I., *Opticks, or a Treatise on the Reflexions, Refractions, Inflexions and Colours of Light* (London: Smith & Walford, 1704).

[12] Photographic lenses became an important side-line of telescope and microscope firms, having distinct technical requirements and much more varied market. Important examples include the companies of Alvan Clark and Sons (1846, USA), telescope-makers; Carl Zeiss (1816–1888, Germany), founded in 1847 for microscope manufacture; John Dallmeyer (1830–1883, England), established in 1859 from within the Andrew Ross telescope manufacturing firm; and Taylor, Taylor and Hobson (1886, England), which focused on scientific instruments and photographic lenses.

[13] See, for example, Abbe, E., 'A contribution to the theory of the microscope and the nature of microscopic vision', *Proceedings of the Bristol Naturalists' Society* 1 (1874): 200–61; Lankford, J., 'Amateurs versus professionals: the controversy over telescope size in late Victorian science', *Isis* 72 (1981): 11–28.

had joined the Zeiss firm in the late 1860s as a consultant researcher, and Hurter founded one of the first research laboratories for the new United Alkali Company in 1891. European chemical companies initially dominated research, with firms such as Belgium's Agfa Gevaert (later important in the photography industry and in manufacturing materials for holography) founding a research laboratory in 1901. British Thomson Houston (UK, 1894)—later the birthplace of the first holograms—followed with their own research arm three years later and, in the USA, General Electric created the first American corporate laboratory in 1900, while Du Pont de Nemours (1903), the National Electric Lamp Association (1908) and the Eastman Kodak Company (1912) established laboratories to develop new chemical and electrical products before the First World War. Like Agfa, Du Pont and Kodak later became important suppliers of materials for holography (hologram research at British Thomson-Houston is further discussed in Section 4.1).

After the war, highly competitive international markets and further mergers encouraged Metropolitan Vickers (UK, 1919) and Imperial Chemical Industries (UK, 1926) to set up their own labs.[14] One source estimates the number of American corporate research laboratories as 300 in 1920 and 1600 just a decade later.[15] Such chemical and electrical firms were to remain at the centre of scientific imaging through the century. The Kodak research lab under Kenneth Mees (1882–1960), for example, rapidly expanded investigations of the type pioneered by Abney, Hurter and Driffield, and became a major producer of scientific research data and standardized procedures in the field. The British photographic firm of Ilford Ltd followed in the 1920s.[16] Such research labs pioneered the introduction of colour photography and spawned further offshoots. The Technicolor Motion Picture Corporation's research lab (1925), for example, was directed by Leonard Troland, a Harvard academic and PhD psychologist with experience at previous industrial labs.[17]

Government laboratories grew alongside their corporate counterparts, but with the distinct aims of quality testing and standardization of optical, lighting and photographic products. The Physikalisch-Technische Reichsanstalt (Germany, 1887), The National Physical Laboratory (UK, 1889) and Bureau of Standards (USA, 1901) set up photometry

[14] Sanderson, M., 'Research and the firm in British industry, 1919–39', *Social Studies of Science* 2 (1972): 107–51; Clayton, R. J. and J. Algar, *The GEC Research Laboratories 1919–1984* (London: Institution of Engineering and Technology, 1989); Reich, L. K., *The Making of Industrial Research, Science and Business at GE and Bell* (New York: Cambridge University Press, 1985); Reader, W. J., *Imperial Chemical Industries: A History* (Oxford: Oxford University Press, 1975); Homburg, E., 'The emergence of research laboratories in the dyestuffs industry 1870–1900', *British Journal for the History of Science* 25 (1992): 91–111.

[15] Dupree, A. H., *Science in the Federal Government: A History of Policies and Activities* (Baltimore: Johns Hopkins University Press, 1986), p.300.

[16] Mees, C. E. K., From Dry Plates to Ektachrome Film: A Story of Photographic Research (New York: Ziff-Davis, 1961); Harrison, G. B., 'The laboratories of Ilford Limited', Proceedings of the Royal Society of London. Series B, Biological Sciences 142 (1954): 9–20. On Mees's notions of research labs as protected spaces for creative individuals, see Shapin, S., The Scientific Life: A Moral History of a Late Modern Vocation (Chicago: University of Chicago Press, 2009), pp 136–57, 183–206.

[17] Southall, J. P. C., 'Leonard Thompson Troland', *Journal of the Optical Society of America* 22 (1932): 509–11.

divisions. The growing culture of corporate and government labs was buttressed by the birth of professional scientific societies relating to imaging, notably the Optical Society of America and the Society of Motion Picture Engineers (1916), the Society of Photographic Science and Technology (Japan, 1926) and the Photographic Society of America (1934).[18]

2.3 THE STEREOSCOPE: SCIENTIFIC MAGIC AT HOME

Scientific imagery transformed Victorian culture. Over the first fifty years after its invention, the consumption of photographic imagery took place principally in the home. Portrait photographs, photographic family albums and the craze for optical toys developed a more discerning market and taste for visual representations. These nineteenth-century cultural experiences, repeated with early twentieth-century visual media, were to underpin the public engagement with holograms a century later.

Daguerreotypes, soon followed by cheaper tintypes and succeeding processes, provided inexpensive individual and family portraits to non-elite audiences for the first time. The mass market potential of middle- and working-class consumption of photographs encouraged an expanding industry. Entrepreneur-photographers opened portrait studios in most towns. Even more popular were locales such as beaches and piers catering to recreation and entertainment. There, bright sun, short exposures and snap decisions by customers combined to yield commercial success.[19]

Audiences quickly grew accustomed to the new portraiture and incorporated it into their home lives.[20] The late Victorian parlour increasingly displayed framed photographs. For those of more limited means, a vogue for collecting small inexpensive photographs expanded. They were a new form of memento collection of the kind that had incorporated autographs, entry tickets, prize ribbons or pamphlets for earlier generations.[21]

[18] As careers in imaging science developed through the century, subsequent professional groupings included the National Association of Photographic Manufacturers (1946), Society of Photographic Engineers (1947), Society of Motion Picture and Television Engineers (1950), Society of Photographic Instrumentation Engineers (1955) and Society of Photographic Scientists and Engineers (1957).

[19] This explosion of enthusiasm for photographs was later seen as a likely model for the progression of holograms in pop culture; see Johnston, S. F., 'Absorbing new subjects: holography as an analog of photography', *Physics in Perspective* 8 (2006): 164–88.

[20] Griffen, M., *Amateur photography and pictorial aesthetics: Influences of organization and industry on cultural production*, PhD thesis, University of Pennsylvania (1987); Griffen, M., 'Between art and industry: amateur photography and middle-brow culture', in: L. Gross (ed.), *On the Margins of Art Worlds* (Boulder, CO: Westview Press, 1995), pp 183–205.

[21] Walker, A. L. and R. K. Moulton, 'Photo albums: images of time and reflections of self', *Qualitative Sociology* 12 (1989): 155–82; Coe, B. and P. Gates, *The Snapshot Photograph: The Rise of Popular Photography, 1888–1939* (London: Ash & Grant, 1977).

Yet mass audiences were fascinated not only with photography, but with other products of scientific imaging. This was an age of visual tricks, when 'philosophical toys' could reveal baffling optical effects. Scientists introduced optical toys such as the kaleidoscope (invented by physicist Sir David Brewster in 1815); the zoetrope (by mathematician William George Horner in 1833) and rapid development of the magic lantern (conceived by Christian Huygens c1650). As holograms would do for a later generation, these devices dazzled the Victorian viewer with vivid coloured shapes, ephemeral animated scenes and spectacular projections, respectively. Thus spectacle, surprise and satiation combined in these novel technologies.

Cinema historian Tom Gunning has argued that such media represented evolutionary steps towards motion pictures, but I claim more generally that they exemplified the cultural links between optical innovation and visual spectacle.[22] The viewing public became increasingly keen for new visual surprises, and the most successful of these scientific inventions was the stereoscope. Scientific experiments to study the visual perception of depth had led Charles Wheatstone to develop the device in 1838. His technique relied on photographing a scene twice, with the camera moved between exposures by the distance between human eyes. These photographs would be recombined visually by using lenses and mirrors to direct the view of each eye to its respective photograph (Box 2.1). A decade later, David Brewster—already known for his invention of the kaleidoscope and other optical investigations—invented a less cumbersome stereoscope that used wedged lenses to divert and focus the eyes towards the two photographic images. When the distinct images were merged by the brain, viewers perceived a three-dimensional representation of the original scene.

Box 2.1 Using stereoscopes

Victorian stereoscopes and their successors present a different photographic image to each eye. Lenses or mirrors overlap the views of each eye to visually 'fuse' these images, yielding a three-dimensional view from a fixed position. Stereo images may be created from any scene that a camera can record from different positions, and range from photomicrographs to three-dimensional images of the moon. Stereo views may be in monochrome or colour, printed on paper or transparencies, or projected on a screen.

The illusion of three-dimensionality is not perfect. The focus of the image is determined by the camera that took the photographs rather than by the viewer's eyes. This fixed-focus characteristic can cause viewing fatigue.

[22] Gunning, T., 'Hand and eye: excavating a new technology of the image in the Victorian era', *Victorian Studies* 54 (2012): 495–516.

As noted by art critic Jonathan Crary, the stereoscope defined one of the principal modes of experiencing photographic images for decades.[23] The popular appeal of the invention spread quickly as a direct result of the first international technology fair, the Great Exhibition of the Works of Industry of all Nations, in 1851, and prototypes of the device manufactured in Paris by Jules Dubosq, an optical instrument maker.[24]

The fashion for stereography also relied on its public supporters: Queen Victoria ordered one, and within five years some half-million stereo views had been produced commercially. In America, physician Oliver Wendell Holmes promoted interest in the new invention, where it had a similarly exponential rise in popularity.

A crucial and unique feature was the visual depth of the images themselves. Viewers quickly discovered that stereography was immersive. Stereoscopes required the user to look into a device that closed off the field of view and replaced the visual world with a new, detailed and deep one. The most popular versions, constructed from steam-shaped wood veneers, form-fitted the face like a scuba mask and focused attention just as the blinkers of a carriage horse did (Figure 2.2).[25] Users could examine and savour

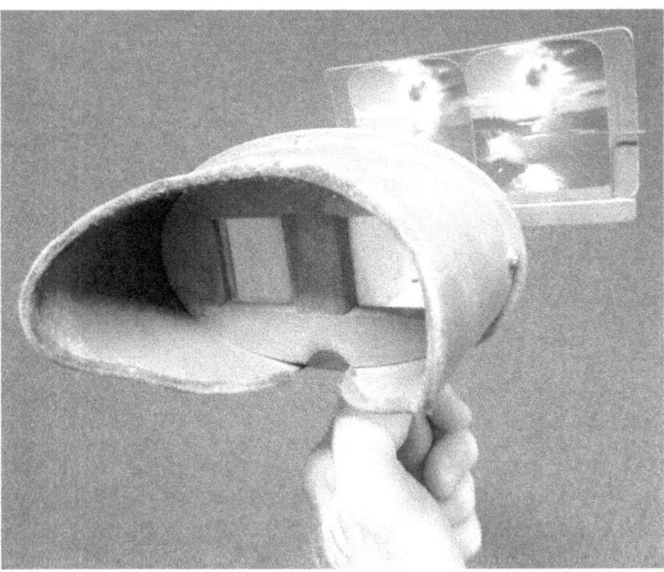

Fig. 2.2 Holmes stereoscope, c1860, with stereo view slide (Young, Y., 'Moonlight, Niagara Falls, N.Y., U.S.A.' (New York: American Stereoscopic Co., 1901); S. Johnston photo).

[23] Crary, J., *Techniques of the Observer: On Vision and Modernity in the Nineteenth Century* (Cambridge, MA: MIT Press, 1990), p. 118.

[24] Brewster, D., *The Stereoscope: Its History, Theory and Construction, with its application to the fine and useful arts and to education* (Hastings on Hudson, NY: Morgan & Morgan, 1856).

[25] Contemporary alternatives were usually table-mounted apparatus or larger pieces of furniture, some of which presented an automated succession of up to 100 views. Such innovation in viewing devices was remarkably similar to entrepreneurs' efforts to market holograms during the 1980s, e.g. Figures 8.4 and 10.2.

fascinating scenes with a directed attention and a sense of vicarious and private—almost voyeuristic—participation. Even monochrome or hand-tinted stereo views provided a sense of realism superior to other visual media.[26]

Visual excitement was at the heart of this scientific device. Stereo views were not glanced at; they were explored to discover unexpected detail, and to experience surprise, fright and even transcendence. As Holmes described,

> The first effect of looking at a good photograph through the stereoscope is a surprise such as no painting ever produced. The mind feels its way into the very depths of the picture. The scraggy branches of a tree in the foreground run out at us as if they would scratch our eyes out. The elbow of a figure stands forth so as to make us almost uncomfortable. Then there is such a frightful amount of detail, that we have the same sense of infinite complexity which Nature gives us.[27]

The experience could be visceral, enthralling and sublime. Views of mountain-scapes, tall buildings and monuments employed that depth of perspective to impress upon viewers a sense of height or grand vistas. More subtly, three-dimensionality encouraged a depth of interpretation as the eye roamed from foreground details to their wider contexts. And given the vividness and depth of imagery, a covert market for erotic images developed steadily.

Rather than recording scenes having personal significance, stereo views captured distant and exciting views for mass consumption. An advantage of such visual topics was their extensibility: new scenes could be discovered or created endlessly by entrepreneurial photographers alert to fresh opportunities. In an age when few Americans travelled except for necessity, even stereographic series of daily life in neighbouring states became popular. During the 1870s a New Hampshire company manufactured some sixty stereo views showing the White Mountain region alone.[28] Other firms produced encyclopaedic collections of natural wonders, national monuments and even stereographic stagings of Biblical scenes. With such graphical 'lessons' illustrating anthropology, sociology, history and geography these offerings were aimed increasingly at primary and Sunday schools as a compelling medium for education.[29]

Stereo views also relied increasingly upon episodic novelty and wonder. Current events were a fertile source of scenes; ceremonies such as the Japanese emperor's birthday, for instance, could show off military parades.[30] Technological accidents such as

[26] On the (imperfect) reality of stereo views, see Crary, J., *Techniques of the Observer: On Vision and Modernity in the Nineteenth Century* (Cambridge, MA: MIT Press, 1990), pp 121–34 and Schroeter, J., *3D: History, Theory and Aesthetics of the Transplane Image* (New York: Bloomsbury Academic, 2014).

[27] Holmes, O. W., 'The stereoscope and the stereograph', *The Atlantic Monthly* III (1859): 738–48.

[28] 'American Scenery, New England Series: stereo view', (c1875).

[29] For such uses, extensive explanatory texts on the back of the stereo view accompanied the captions. The products of Underwood & Underwood Publishers ('New York, London, Toronto and Ottawa, KS') typically incorporated 350-word texts. Dates, on the other hand, seldom appeared, in order to preserve the currency of the image and its continued potential for generating sales.

[30] '2331. Great Military Review on Emperor's Birthday, His Majesty attending, Qoyama Parade Grounds, Tokio, Japan: stereo view', (H. C. White Co, 1905).

train wrecks became an important subgenre, the visual aftermath of conflagrations, earthquakes and typhoons proved popular and military scenes had a similar fascination. The upsurge of stereoscopes was ideally timed for vicarious experience of the American Civil War.

The business of stereographic views also encouraged an international market and production system. The most easily captured scenes—exotic dress, scenic locales and impressive architecture—appealed most to viewers in other countries unable to visit them. Photographers commonly licensed their images to publishers in other countries to supply home and international markets, an economic arrangement much later reproduced in news photography and commercial holography. Spectacular imagery became generic and universally appealing, and cottage businesses rapidly scaled up. Relying on hand-printing original photographs during the first decades of production, factories of workers could chemically process, cut, glue, package and distribute the stereo views at an affordable price.

The commercial market for stereograms rapidly assumed international dimensions: the unfamiliarity of far-away places proved captivating, and photographers discovered an exploding global market for their easily-transported views. The products of the Strohmeyer & Wyman Company of New York, for example, were printed with captions in six languages. The Pettijohn Company labelled its views with the catchphrase, 'Around the world by stereoscope. Without leaving your home—just like being there'.[31] And Underwood & Underwood marketed a 'Stereograph Travel System' consisting of hundreds of stereo views indexed by maps and locations, designed to be an alternative to the expensive world tour for aspiring middle-class purchasers.

Stereoscopes, then, offered rich opportunities for visual entertainment and education. Their popularity was strong from the 1850s up to the turn of the century as new content expanded to reach new audiences.

Visual interest proved difficult to sustain into the twentieth century, however. The last technological innovation to sustain markets was the introduction of printing processes that could reproduce photographic images, offering lower costs along with the exciting introduction of colour scenes. The falling price of stereo views themselves also arguably made them less appealing.[32] Just as the introduction of encyclopaedias on CD-ROM during the 1980s, and later free-access internet sites a decade later, devalued the appeal of printed encyclopaedias, cheaply printed stereo views made them ubiquitous, eroding the sense of exclusiveness felt by middle-class audiences two generations earlier. As the cycles of appeal of twenty-first-century products reveal, mass markets often begin with elite purchasers and end when everyone can, and does, have one.

[31] 'Going to church in Sweden: stereo view', (Pettijohn Packages, 1906).

[32] Monochrome stereo views cost about a dollar per half-dozen in the 1860s, the equivalent of some $4 US each in 2015 currency. By 1910, with the conversion of the industry to colour lithography and half-tone reproduction, equivalent prices dropped by about 80% (DeLeskie, P., *The Underwood Stereograph Travel System: a Historical and Cultural Analysis*, MA thesis, Concordia University (2001)).

Propriety and a growing public association of stereo views with pornography also had an influence on their decline—another recurring theme in visual media. As an early motion picture promoter recalled decades later,

> We remembered hearing what happened to the stereoscope in the days when our fathers were young. That very attractive instrument, showing beautiful scenery in natural deep relief, was to be found in nearly every lady's drawing-room, until in an evil day some unprincipled persons began selling indecent photographs for use in it. That was its knell. It speedily acquired such ill-repute that it was totally banished and never came back into favour.[33]

And finally, competition played a role. New media challenged the supremacy of the stereoscope. Picture periodicals (discussed in Chapter 3) exploded from the turn of the century, offering the consumption of imagery more cheaply and quickly. Stereography, ubiquitous for two generations, was no longer exhilarating. Visual thrills required constant stimulation and extension, and the stereoscope had used up its stock of visual surprise and technical innovation. The life story of the stereoscope had remarkable parallels with the experiences of holograms a century later.

2.4 HALTING SCIENCE: THE CASE OF THE LIPPMANN PHOTOGRAPH

These sketches of technical change and cultural adoption suggest progressive expansion and improvement and scarcely hint at popular critiques. Popular resistance becomes more evident if we focus on cases of overt failure rather than success: as the innovators appreciated, not all scientific developments in imaging processes were seen as advances by potential consumers. The critical evaluation of processes for colour photography illustrates the community criteria applied to imaging technologies.

The enthusiast–company research axis was a category of activity dubbed 'applied science' in the early twentieth century. Countering it was academic or 'pure' science, which relied on a growing body of theory as well as experimental investigations. Such 'pure' pursuit of intellectual curiosity had led quickly to the 'applied' commercial successes of optical toys a generation earlier, but sophisticated science could as easily be stopped in its tracks by critical audiences.

The turn-of-the-century exemplar of theory-led optical science was Gabriel Lippmann (1845–1921), a French physicist who devoted much of his career to photography and other imaging processes. During the 1890s he invented a wholly novel method of recording colour images based on the interference of light. The technique and its mathematical explanation won him one of the first Nobel Prizes in Physics, and proved to be a tantalizing precursor to holograms developed some seventy years later (discussed in

[33] Hepworth, C. M., *Came the Dawn: Memories of a Film Pioneer* (London: Phoenix House, 1951), quotation p. 108.

Section 4.4).[34] And—again hinting at the later attractions of holograms—Lippmann also explored means for recording fully three-dimensional images that did not require special viewing apparatus.[35] Just as relevantly, both Lippmann's spectacular inventions found diffident support from peers, enthusiasts and the wider public.

A combination of technical and social factors contributed to the repeated rejections of early schemes of colour photography. The need for special apparatus was a recurring theme, but was not in itself the crux of the problem. Colour photographic processes through the early twentieth century required arcane set-ups or special commercial emulsions. Three-colour photography was particularly awkward. It relied on separate exposures of three photographic plates—one each through a red, a green and a blue filter. Each plate, combined with its respective colour filter, would then be projected onto a screen to yield an overlapping full-colour image. Alternatively, the three images could be printed in overlapping coloured inks, or combined in an assembly that incorporated colour filters for viewing. A simpler but still expensive technique for amateurs was based on miniature screens placed over the emulsion or incorporated into the emulsion to filter light into three primary colours. The most successful of these was the 1907 *Autochrome* process of the Lumière brothers in France, arguably the most innovative and prolific developers of imaging processes. Their method relied on intermingling grains of starch, dyed in three primary colours, overlying a standard photographic emulsion. As long as these integral filter screens remained accurately registered between exposure and chemical processing, a good quality full-colour transparency was obtained.

The technical characteristics shaped cultural uptake. Each of these colour processes was at least an order of magnitude more expensive than regular black-and-white photography. And in each case, the need to use uncommon panchromatic emulsions sensitive to the three primary colours, along with the dense colour filters themselves, resulted in much longer exposure times. The subject matter of photographs was consequently limited. As with the slow emulsions used in earlier photographic processes, motion and candid scenes were difficult to record. And the screen-process techniques, by dividing up the emulsion surface into stripes or grains, inevitably reduced the resolution of the resulting photographs. Spontaneity and detail were sacrificed for this new technical dimension, and amateurs widely judged this to be an unappealing trade-off.

Gabriel Lippmann's *interferential* or *interference colour photography* was a fascinating but unpromising competitor for such jostling techniques. Lippmann, Professor of

[34] Lippmann, G., 'La photographie des couleurs', *Comptes Rendus de L'Academie des Sciences* 112 (1891): 274; Lippmann, G., 'Sur la théorie de la photographie des couleurs simples et composés', *Journal de physique* 3-e serie 3 (1894): 97–107; Lippmann, G., 'Colour photography', *1908 Nobel Prize Lecture* (1908); see also Connes, P., 'Silver salts and standing waves: the history of interference colour photography', *Journal of Optics* 18 (1987): 147–66. A detailed survey of the intensive development of colour photography processes is given in Friedman, J. S., *History of Color Photography* (London: Focal Press, 1972).

[35] Lippmann, G., 'Épreuves réversibles. Photographies intégrales', *Comptes Rendus de l'Académie des Sciences*. 146 (1908): 446–51. For an overview of three-dimensional imaging technologies, see the Appendix. See also Phillips, N. J., H. Heyworth and T. Hare, 'On Lippmann's photography', *Journal of Photographic Science* 32 (1984): 158–69.

Experimental Physics at the Paris Academy of Science from the late 1880s, had laboriously developed a proof-of-concept technique but not a commercial process.

His sophisticated colour photography had characteristics superficially like the other examples of late-Victorian photography: it relied on a standard camera, but required a special plate holder and very special photographic plate. A complication was that the plate holder was required to hold the emulsion in contact with a shallow pool of mercury, which acted as a form-fitting mirror. The scheme ensured that light focused by the camera lens onto the emulsion would also reflect from the mercury interface and travel back into the emulsion again.

The basis of Lippmann's invention relied on high science: contemporary experiments were confirming that light could be understood as a wave. His apparatus would form a standing wave, with peaks of intensity every half-wavelength through the thick gelatine emulsion. After development, these intensity variations would be recorded as planes of silver salts. When lit with full-spectrum (white) light such as sunlight, the set of reflective planes would reinforce and reflect light waves of the colour (wavelength) originally used to record them. Even less intuitively for contemporary observers, a Lippmann colour photograph could also record images of mixed colour. The colour palette, consisting of a collection of wavelengths, would produce a more complex standing wave that reflected the combination faithfully in the right circumstances. This was the highest of high science: Lippmann explained it through the mathematics of Fourier decomposition, a topic then unfamiliar to commercial technologists and many practising scientists.

There were practical complications, too. The technique invited superlatives. The plate itself was coated with a remarkably fine-grained emulsion required to record the variations in intensity occurring over the space of a fraction of a wavelength of light. This was a chemical tour-de-force that was far more technically demanding than the Lumières' subsequent coloured-screen techniques. To record intensity variations or 'fringes' of violet light, for example, a spatial resolution of the order of a hundred-millionth of a metre was needed, hundreds of times better than then-available films. The emulsion had a sensitivity to light ('speed') some one- to ten-thousand times less than regular emulsions.

Yet the science itself was a first point of attack. French colleagues criticized the seeming arbitrariness in the chemical processes that sensitized and stabilized Lippmann's photographic emulsion. For these technical experts, Lippmann's interferential photography was neither, as supporters described it, 'direct photography', 'natural colour' nor 'objective colour reproduction'. Instead of being a technique that mapped a physical attribute unambiguously onto a material object, it was just as capricious and subjective as three-colour methods of generating colour photographs. To balance emulsion sensitivity for different colours, Lippmann's demonstrations relied on special dyes to absorb more light and to make the emulsion preferentially sensitive. And, to further compensate the colour rendition, he adjusted the red, blue and green components in his exposures to compensate for the emulsion's unequal sensitivity to different parts of the spectrum. His expert critics argued that the elegant physical principles of Lippmann's method of photography

Box 2.2 Viewing Lippmann photographs

As with other contemporary colour processes (and later holograms), viewing a Lippmann photograph was a vexing procedure. The plate has to be lit by diffuse unfocused light, and preferably natural skylight entering an otherwise darkened room, rather than by the impoverished yellow-tinted light of gas or electric lamps. The plate must also be lit and viewed nearly face-on so that the optical interference in the emulsion yields the correct colour (viewing at an angle shifts the colours towards the blue end of the spectrum). To avoid the viewer blocking the source of light or being dazzled by the reflection from the glass surface, this often meant attaching a shallow glass prism to the plate so that light, eyes and plate could be arranged appropriately. Projecting the images was difficult at best, and they could not be copied by the usual photographic method of contact exposures.[36]

were mired in the messy chemistry of emulsions and the practical engineering details needed to obtain and view images (Box 2.2).[37]

As summarized by a present-day Lippmann experimentalist,

> The viewing of Lippmann photographs has to be carefully managed in order to see any colours. Under relaxed viewing conditions provided by a large diffuse source, problems arise with colour reconstruction. At very close quarters you can see colour shifts across the width of the plate. Colour shift is quite noticeable within 5–10 degrees, and as you walk past the plate the colour shift is extremely disturbing! If a group of people gather round a Lippmann plate, those who can see any colours will see different ones to their neighbour. In the past they had projectors that allowed several spectators to view a Lippmann plate simultaneously, and all would see the same colours; whether these were the correct colours is another matter. So long as trees looked green and red bricks looked red most observers would be happy.[38]

This hidden subjectivity—not only of technical constraints, but also of contemporary proponents' reticence to communicate them—inhibited scientific engagement. Lippmann's

[36] Bjelkhagen, H. I., 'Lippmann photography: its history and recent development', *The PhotoHistorian* 141–2 (2003): 11–9; Friedman, J. S., *History of Color Photography* (London: Focal Press, 1972), pp 23–33.

[37] Mitchell, D. J., 'Reflecting nature: chemistry and comprehensibility in Gabriel Lippmann's "physical" method of photographing colours', *Notes and Records of the Royal Society* 64 (2010): 319–37; McHelone, J. P., 'The signature of light: photo-sensitive materials in the nineteenth century', in: A. Thomas, M. Braun and National Gallery of Canada (eds.), *Beauty of Another Order: Photography in Science* (New Haven, CT: Yale University Press, 1997), pp 60–75. The Lumière brothers developed the technique to further tame the experimental complexity and to make the emulsion less sensitive to swelling and colour distortion, but had little commercial success.

[38] Handheld viewers and the Zeiss aphengescope projector effectively 'forced optimum viewing conditions upon unsuspecting spectators . . . this technical limitation was concealed from viewers and it would not have been a very good selling point' (Green, D. to SFJ, emails, 7–14 Aug 2014, London, SFJ collection). Similar issues were later at play in viewing colour holograms.

system, while eventually garnering the 1908 Nobel Prize, failed to satisfy the reasonable expectations of contemporary professional and popular cultures. For wider publics, criticisms about theoretical elegance were unimportant; what mattered was the practicality of Lippmann's technique for contemporary consumers. His emulsions, much slower than other all-colour (panchromatic) emulsions of the time, meant that still-life subjects or posed portraits of Daguerrean rigour were the only options for photographic enthusiasts.

These were not technical objections alone; as subsequent examples will show, imaging processes have always constrained the nature of the images they can record. Fifty years earlier, Daguerre's results had unlocked doors, exciting audiences having few expectations and liberating them to explore the novel applications of photographs. What made colour processes, and particularly Lippmann's technique, unacceptable was that they mapped so poorly onto contemporary cultural requirements. In an age transformed by mechanization and speed, time-exposures could no longer provide a meaningful representation of the modern world.[39]

By the early twentieth century, then, the refinement of empirical understandings of photographic chemicals and materials, the mathematics of optics and the systematization of research were creating communities that were establishing higher standards and a professional culture of scientific imaging. But along with them, popular audiences were gaining an appetite for more spectacular extensions to visual culture.

[39] An example of the disdain of Lippmann's contemporaries can be found in 'Editorial', *British Journal of Photography*, 38, 20 Feb 1891: 17.

3

Grassroots Modernity

Holograms did not appear out of the blue. They developed against a changing cultural backdrop that increasingly melded imagery with science. As introduced in Chapter 2, the run-up to their emergence involved decades of innovation and reconception that gave technical professionals, enthusiasts and the public alike a taste for scientific framing, visual novelty and perpetual innovation. To understand the contexts and impacts of popular culture, this chapter carries forward the discussion to the half-century that preceded holograms and the imaging technologies that captivated earlier audiences. It extends the claim that audiences for holograms were acclimatized by preceding technologies and cultural contexts.

The rising scientific culture of the early twentieth century underpinned growing popular engagement with imaging. At least three routes contributed to this diffusion: the availability of new technical processes for image reproduction, a proliferating variety of startling images for mass consumption in the print media and the extension of imaging technologies to amateur scientific enthusiasts. Together, they promoted a rising visual literacy and appetite in wider publics. To an unprecedented degree, imagery became both democratic and startling.

3.1 REPRODUCTION FOR THE MILLIONS

Just as chemical imaging was being made more scientifically rigorous, practical as an engineering art and attractive as a popular medium, *mechanical* imaging, too, was being improved.

The reproduction of images was an ancient art, but had been refined during the nineteenth century. The printing plate could be a wood-cut, engraved metal plate or (by the late eighteenth century) an etched stone surface (lithograph); ink was held on the uncut portions of the wood or metal, or un-etched portions of the stone. The art of engraving and lithography could nevertheless reproduce different tones of grey by fine cross-hatching of varying density and style.[1]

[1] Meggs, P. B., *A History of Graphic Design* (New York: John Wiley & Sons, 1998); Moran, J., *Printing Presses: History and Development from the Fifteenth Century to Modern Times* (Berkeley: University of California, 1973).

During the late nineteenth century these basic approaches were developed in a direction that could be described as an early form of digitization. The *halftone* method represents different tones of grey via dots of ink of different size or density. The method relies on first re-photographing the desired image with a fine screen over the emulsion to divide it into individual unfocused cells, each of which will have an intense centre and dimmer edges. When the emulsion is then developed to yield a high-contrast negative, the result will be a fine pattern of spots whose sizes depend on the original intensity. This binary pattern can be reproduced on a printing plate and then printed onto paper pages. Perfected in the early 1890s, the halftone screen process mechanized image reproduction; the cost and time needed to reproduce images plummeted, and their content moved from artistic to photographic.

The invention of the halftone process consequently produced a step-change in public access to photographs. Instead of being limited to a small collection of personal tintype portraits or handmade stereoscope views, fin de siècle audiences could view photographic images in a growing range of newspapers, magazines and cheap stereo views. Print media began a rapid cultural transition towards photographic imagery.[2]

3.2 IMAGERY TO EXHILARATE

> *Technology has subjected the human sensorium to a complex kind of training. There came a day when a new and urgent need for stimuli was met by the film. In a film, a perception in the form of shocks was established as a formal principle.*[3]
>
> Walter Benjamin, 1940

Combined with inexpensive reproduction in periodicals, photography transformed visual experience for twentieth-century publics. In the hands of photographers and editors, the content of that imagery was increasingly designed to startle and disorient. Audiences for photography, cinema and contemporary art stumbled through jarringly different expressive forms through the first three decades of the century. High art was reproduced for the millions in large-circulation magazines and quickly popularized in new forms that had even wider appeal. Modernity was portrayed through exhilarating graphics, and imagery became a key element of the modern world.

[2] Phillips, D., *Art for Industry's Sake: Halftone Technology, Mass Photography, and the Social Transformation of American Print Culture 1880–1920*, PhD thesis, Yale University (1996); Jussim, E., *Visual Communication and the Graphic Arts: Photographic Technologies in the Nineteenth Century* (New York: R. R. Bowker, 1974). This new cultural emphasis on images also provided a resource for historians to better understand the rapidly developing visual culture at the turn of the century onwards (Tucker, J., 'The historian, the picture, and the archive', *Isis* 97 (2006): 111–20).

[3] Benjamin, W. and H. Zohn (transl.), 'On some motifs in Baudelaire', in: H. Arendt (ed.), *Illuminations: Essays and Reflections* (New York: Schocken, 1968), pp 155–200; quotation p.175. See also Gunning, T., 'The cinema of attraction: early film, its spectator, and the Avant-Garde', *Wide Angle* 8 (1986): 63–70.

Visual ideas introduced by a handful of artists spread internationally with unprecedented speed. Among the most influential was Alexander Rodchenko (1891–1956), a seminal creator of sculptural artworks, photographs and graphic art in the young Soviet Union. His contributions to photography were arguably his most influential work. Rodchenko and his peers in the Soviet Constructivist movement pioneered ideas that were rapidly taken up in graphic arts around the world.

The first influential idea was Rodchenko's exploration of unusual perspectives. Rodchenko's work countered the turn-of-the-century *pictorialist* movement, which had sought to turn the craft of photography into an art form modelled on the aesthetic principles of drawing and painting. Artistic notions of perspective, composition and portrayal of reflective mood had been central to its practitioners.[4] This competition between spectacle and disorientation, on the one hand, and traditional aesthetics, on the other, were to inform later appreciations of holograms.

Photographic conventions had been shaped by both culture and technology. In some respects, camera design until the first decade of the twentieth century had also constrained visual ideas. Camera designers sought faithful reproduction of conventional scenes, and their criteria were shaped by contemporary art and urban culture; their cameras consequently embodied and facilitated this aesthetic. As a result, serious photographers had been materially encumbered and perceptually channelled by large-format, tripod-mounted plate cameras. The unwieldy hardware was designed to reproduce the horizontal view of human eyes as well as the traditional perspective of paintings. The designs of equipment consequently favoured these conventions. Incorporating adjustments to tilt and offset their lenses, so-called 'view cameras' were used to flatten imagery, remove vertical perspective and keep all parts of the scene in focus. Pointing upwards or downwards at steep angles was further discouraged by problems of camera stability for the long exposures typically used and by the difficulty of loading photographic plates and viewing the ground-glass to focus and compose the picture. As a result, good photographic images aimed to be horizontal, well-focused and carefully posed.

Rodchenko, by contrast, adopted compact and hand-held cameras and pointed them in unfamiliar directions. This was both a technological and cultural shift. To gaze sharply upward or downward was (and still is, in some settings) an unusual social behaviour.

[4] Pictorialism, as an early engagement between the world of fine art and imaging technologies, foreshadowed later art-cultural critiques of holograms. Key proponents of pictorialism included Henry Peach Robinson (1830–1901, Britain) and Alfred Steiglitz (1864–1946, USA), although this aesthetic of art photography also became established in several European countries and Japan from the 1890s. The Linked Ring Brotherhood, founded by Robinson in Britain in 1892, promoted photography as a new variety of art requiring high technical expertise and aesthetic refinement. Its members withdrew from what they saw as the staidly academic and overly-inclusive Royal Photographic Society. Its American counterpart was the Photo-Secession movement, championed from 1902 by Alfred Steiglitz via a series of organizations, magazines and galleries. Pictorialism signalled a new generation of serious amateur photographers, and its influence endured generations beyond them in photographic societies and camera club exhibitions through the twentieth century. For its origins, see Robinson, H. P., *Pictorial Effect in Photography: Being Hints on Composition and Chiaroscuro for Photographers* (London: Piper & Carter, 1869); Steiglitz, Alfred, *Camera Work* magazine (1903–1917).

Familiar to children, it is culturally discouraged for adults, for whom it may be interpreted as threatening, provocative or naïve. To stare upwards at city high-rise towers is often the mark of a tourist; to peer down from a high vantage point may be labelled as furtive or voyeuristic. And close observation of strangers in public—especially when recorded candidly via a camera for subsequent examination—can be seen as impolite and intimidating. The unconventional pointing of a camera, then, could recover childlike dimensions and fresh perspectives, both visual and social. Rodchenko's photographs defamiliarized imagery (Figure 3.1). The viewer was challenged to recognize and accommodate to the odd perspectives. The result of this perspective-play could be liberating, revealing unnoticed details and viewpoints. His graphic technique was a new form of spectacle that encouraged viewers to appreciate the social and even political dimensions of images. It contrasted the introspective tranquillity of pictorialism with an aggressive celebration of the discordant modern social world. Realistic imagery could now be unnerving and even shocking.[5]

Fig. 3.1 Perspective play: steel vat photographed in the style of Alexander Rodchenko (S. Johnston photo).

[5] On a different variety of surprise—concerning the acceptance of new styles of art by American elites rather than mass audiences—see Hughes, R., *The Shock of the New: Art and the Century of Change* (New York: Thames and Hudson, 1991) and Kammen, M., *Visual Shock: A History of Art Controversies in American Culture* (New York: Random House, 2006). This high culture/low culture divide is discussed in relation to holograms in Section 9.4.

For Rodchenko, this was a carefully theorized strategy:

> Photography—the new, rapid, real presenter of the world—with all its possibilities, would take up the work of showing the world from all points, of educating the ability to see from all sides . . . But the opportunity exists to show the object from the sorts of view-points that we look from but don't see . . . in order to teach man to see from new viewpoints, it is necessary to photograph ordinary, well-known objects from completely unexpected viewpoints . . . We must revolutionize our visual thinking.[6]

This conspicuous suggestion of the importance of point of view highlighted the role of the viewer of the photograph. The photographer's viewpoint was designed to discomfit, and to make the observer complicit in an observation that could be furtive or transgressive. The viewer, no longer a passive recipient, was enrolled as an active participant in interpreting the image—an empowerment having political dimensions.

Rodchenko's images defined and celebrated modernity. His aerial views revealed the machine-like repetition of ranks of marching soldiers and the aerodynamics of a diving bather. His images of modern architecture and especially industrial plants discovered unfamiliar beauty and power in their forms. For the new Soviet state, this juxtaposition of social will with modern technology expressed propagandistic ideals; similar themes, however, were soon popular in interwar Germany and America, even if the political dimensions were differently configured.

This avant-garde style of spectacular imagery, especially when applied to industrial subjects, illustrates historian David Nye's notion of the 'technological sublime'.[7] The sublime is an experience traditionally associated with a supreme being and which transcends the ordinary. It characteristically evokes exaltation, wonder and fear. The technological sublime is such an experience instantiated in a modern technology and, as I argue here, imagery that highlights it.

The second unsettling technique popularized by Rodchenko and his peers—and again mapped onto later holograms in novel ways—was photomontage. The use of photographs as graphic elements in making a collage appears to have begun with the *Dada* movement of art in Germany immediately after the First World War, and the technique was rapidly taken up in the young USSR by Rodchenko and other members of the Constructivist art movement there. By juxtaposing portions of photographs, printed texts, coloured graphical elements and drawings, the Dadaists and Constructivists sought to challenge aesthetic and cultural complacency with 'anti-art'. Their creators argued that, in photomontages, the photograph could shock the viewer much more effectively than artistic creations could, because its documentary quality emphasized the reality and urgency of a visual event. In the hands of the Russian Constructivists such as Rodchenko, Gustav Klutsis and El Lissitzky, photomontages were turned into explicitly political

[6] Rodchenko, A., 'Response to Boris Kushner letter in *Sovetsko foto*, *Novyi LEF* no. 10, 18 Aug 1928', in: J. Gambrell (transl.) and A. N. Lavrentiev (ed.), *Aleksandr Rodchenko: Experiments for the Future* (New York: Museum of Modern Art, 2005), pp 207–12, quotations pp 208, 211, 212.

[7] Nye, D. E., *American Technological Sublime* (Cambridge, MA: MIT Press, 1994).

posters, often combining photographs of crowds, architecture and contemporary leaders such as Lenin and, later, Stalin, in discordant but rousing perspectives. The photograph became part of activist art with the goal of transforming society and culture.[8] Borrowing from Cubist art, the combination of visual elements confused viewers' perceptions of point of view and depth. Like later double-exposure and multi-channel holograms, the photomontage was visually dissonant, combining disturbing realism with the jarring perception of multiple perspectives.[9]

Photojournalism was transformed by this new photographic art. Dissemination in printed mass media such as posters, magazines and newspapers, the expressive power of European photography after the First World War seduced a generation.

New technology, in the form of small and inconspicuous cameras, allowed photographs to be 'captured' in candid and unfamiliar circumstances using available light; the morphology of industrial architecture and the transitory nature of crowds could now be examined at leisure. The new field of photojournalism embraced these graphic approaches and technological tools. Magazines between the world wars were flooded with printed images, and cross-talk between them generated increasingly dramatic and breath-taking views that sought to disorient and unsettle their viewers. Soviet magazines such as *Kino-Fot* (USSR, 1922) and *LEF* (USSR, 1923) were among the first, quickly followed by *Broom* (Italy, 1924), *Arbeiter-Illustrierte-Zeitung (AIZ*, Germany, 1924) and *Sovetskoe foto* (USSR, 1926). A growing number of topical magazines such as *Vu* (France, 1928), *Regards* (France, 1932), *Life* (USA, 1936), *Look* (USA, 1937) and *Picture Post* (UK, 1938) rehearsed and further popularized these visual themes. And what had begun as an explicit movement in avant-garde art and a medium for propaganda grew in the west to inform photoreportage, advertising, interior design and popular culture.[10]

For European and American audiences alike, the common factor was the mass exposure to a new visual grammar. For Russian artists like Rodchenko and Klutsis, photography during the 1920s expressed the popular post-revolutionary slogan, 'Art to the Masses'.[11] During the late twenties, photojournalists formed photographic workshops in the October Association of Artistic Labour with the express purpose of replacing 'picturesque'

[8] Lodder, C., *Russian Constructivism* (New Haven: Yale University Press, 1983), pp 186–204; Tupitsyn, M., *The Soviet Photograph, 1924–1937* (New Haven, CT: Yale University Press, 1996). On the political underpinnings of early Soviet photography, see Cohen, A. J., *Imagining the Unimaginable: World War, Modern Art, and the Politics of Public Culture in Russia, 1914–1917* (Lincoln, NE: University of Nebraska Press, 2008).

[9] Holograms were to rehabilitate photomontage in two ways: first, by allowing overlapping three-dimensional scenes recorded from different perspectives (double-exposure holograms), and second, by presenting scenes that changed with viewing or lighting position (multiple-channel holograms).

[10] See, for example, Lucaites, J. L. and R. Hariman, 'Visual rhetoric, photojournalism, and democratic public culture', *Rhetoric Review* 20 (2001): 37–42.

[11] During the late 1920s another slogan, 'Art of the Masses', described the aim of integrating amateur activities into professional art and graphic design. See Tupitsyn, pp 100–3. This theme later underlay the attractions of counterculture holograms (Chapter 8).

photography with a new socialist visual culture. They found, however, increasing resistance within even the Soviet press to this new form of representation:

> Many photojournalists who submit vivid snapshots experience complete disappointment when the editors do not grasp their 'points of view'. Often [the editors] simply do not understand how a photograph can be tilted or, simply said, 'fall,' or how one can publish a photograph with a close-up, for example, of the details of machines, the movements of hands, etc . . . Above all, editors approve of photographs in which all the elements are fitted into absolutely concrete and intelligible forms for the reader.[12]

3.3 IMAGERY TO EDUCATE

Advertisers began to recognize the potential of such imagery to educate audiences and to influence consumers. Photographs became more frequent in advertising from the turn of the century, and especially through the First World War. This commercial form of 'education' proved effective, but not guaranteed: advertisers found themselves adapting on the fly to changing public perceptions of their images. Straightforward photographs that had captured the attention of early viewers were soon superseded by images designed to focus attention, accentuate product features and to transmit less conscious symbolic messages. This increasing visual sophistication influenced deeper-seated cultural attitudes about authenticity, art and, eventually, persuasive advertising.[13]

Commercial promotion also shored up and rehabilitated visual technologies for new markets. As noted in Section 2.3, education via the stereoscope had been suggested as early as the 1850s. Stereo view publishers aggressively promoted the use of stereoscopes in schools. Visual education became a theme of educationalists during the 1920s, and maps, illustrations, diagrams, graphs and projected images were marshalled as their modern tools of instruction.[14] The teaching of geography was a particular beneficiary, with magazines such as *National Geographic Magazine* (founded in 1888 and adopting full-page photographs in 1905) providing illustrations for classroom use. But as one teacher underlined,

> The finest service yet rendered in the schoolroom has been done by the stereograph. The photograph presents but two dimensions. But the stereo camera and the stereoscope work a miracle. They supply the actuality of binocular vision, and the third dimension is presented to the eye in vivid reality. The person who looks through the stereoscope looks upon the real

[12] Shaikhet, A., 'Sorevnovanie foto-reporterov razvertyvaetsia [The competition of the photojournalists unfolds]', *Sovetskoe foto* (1929): 713, quoted in Tupitsyn, p.102.

[13] Brown, E. H., 'Rationalizing consumption: Lejaren á Hiller and the origins of American advertising photography, 1913–1924', *Enterprise and Society* 1 (2000): 715–38.

[14] See, for example, Brewster, D., *The Stereoscope: Its History, Theory and Construction, with its application to the fine and useful arts and to education* (Hastings on Hudson, NY: Morgan & Morgan, 1856); 'The stereoscope in the schoolroom', *Pennsylvania School Journal* 17 (1868): 150–2; Keystone View Company Education Department, *Visual Education: Teacher's guide to the Keystone '600 Set'* (Meadville, PA: Keystone View Company 1920); Bak, M. A., 'Democracy and discipline: Object lessons and the stereoscope in American education, 1870–1920', *Early Popular Visual Culture* 10 (2012): 147–67.

mountain, looks into the depths of the real canyon, looks upon the actual statue, the actual cathedral . . . It becomes a game to see who can stand and report in good English what he saw, looking through the window of the stereoscope into the reality beyond.[15]

As magazine photographs, early cinema and hobby photography each expanded, visual technologies combined to further influence culture. Photographic advertising could, for instance, illustrate the stereoscope's educational applications more convincingly than the vague claims about the benefits of visual encyclopaedias had done. Placement of advertisements in specialist journals, such as those read by teachers and doctors, targeted the message at relatively large audiences.

Between the world wars, the Keystone View Company, founded in Pennsylvania during the 1890s, was a regular advertiser in such journals and supplied stereoscopes for the classroom along with stereo view collections and educational manual for teachers. Interwar readers were coached to provide their pupils with 'those marvellous third-dimension near experiences' and access to the latest current events.[16]

And by the early 1940s, as wartime urgency underlined the need to train children and adults efficiently, the company refined its messages in *The Educational Screen: the Magazine Devoted Exclusively to the Visual Idea in Education*:

> If the use of the stereoscope and stereoscopic pictures is so important to the efficiency of the Army and Navy and to the successful conduct of the war, it seems hardly necessary to emphasize its importance in the education of our children for their future—whether the future involves war or peace. With its impressive elements of reality, the stereograph brings both to the soldier and to the student factual information that can be obtained in no other way.[17]

Stereo views were also reconceived as high-impact illustrations in books and magazines by the provision of a folding portable viewer attached to back covers (Figure 3.2).

Three-dimensionality was reimagined as a potent advertising medium, providing a better way of communicating product features in travelling sales presentations. View-Master marketed their updated stereoscopes and 'reels' of seven stereo views 'to add color and depth to sales stories' as well as children's toys.[18] But educational and promotional

[15] Good, J. P., 'The scope and outlook of visual education', *School Science and Mathematics* 20 (1920): 482–7, quotation p. 483. The author noted, however, that cinema could do even more: 'The stereograph arrives at perfection, in presenting the perception of solidity, and distance, the third dimension of the view. There is nothing to compare with it in this service, but it is a static world. Motion is absent. Yet motion is another "dimension," and the presentation of motion in the picture is an arrival at another apex of perfection. The gracefully moving animal, the rushing waves, the swaying trees, are all there, to the last perfect detail of motion' (p. 484). A contemporary argued that cinema was a superior medium because it allowed an entire class to be instructed instead of usage one by one, which wasted valuable time (von Hentig, H., 'Education and the cinematograph', *Moving Picture World* 10 (1911): 973–4).

[16] Keystone View Company, p. a., 'Oliver Wendell Holmes was right: the stereoscope is not a toy', *The Educational Screen: the Magazine Devoted Exclusively to the Visual Idea in Education* 16 (1937): 129.

[17] Keystone View Company, p. a., 'The stereoscope goes to war', *The Educational Screen: the Magazine Devoted Exclusively to the Visual Idea in Education* 16 (1942): 275.

[18] Sawyer's Inc, advertisement, 'Sell your products with View-Master', *Business Screen* 10, 11 (1949–50): 8.

Fig. 3.2 Stereoscopy for publications. Stereo photographs combined with fold-out metal binocular viewer (Farrar & Rinehart Stereo-Books, New York, 1937; S. Johnston collection).

usages were, at best, supplements to existing markets and tempted relatively few mid-century schools or firms. Each of these marketing arguments and outcomes was to be rehearsed decades later for holograms.

Soviet photography had sought to shake up and educate the masses—to literally and figuratively shift perspectives—with utilitarian art, but a growing Soviet orthodoxy had limited its growth. Photojournalism in the West sought something similar, if only to sell magazines and entertain consumers with time on their hands. The mass market for such magazines and new imagery expanded rapidly. An example of this democratization is the spreading of modern imagery in even wide-circulation magazines aimed consciously at 'everyman' and 'everywoman'. During the 1930s and 1940s, magazine readers around the western world were presented with this new genre of disorienting and surprising photographs (Figure 3.3).

Wonder could be elicited not just by viewing the world from unconventional angles, but also by viewing the ultra-small or invisible (Figure 3.4). Amateurs were similarly encouraged to add excitement to their photographs by tricks made possible with optical and mechanical tinkering.[19]

These three varieties of spectacular imagery—ranging from early Soviet photographic art as education and ideology, to photojournalistic representations of sublime

[19] See, for example, 'Taking unusual photos with pinhole cameras', *Popular Mechanics*, 60 (3), Sep 1933: 428–9; 'Stunts with your camera', *Popular Mechanics*, 59 (6), Jun 1933: 874–8; 'The world under a lens', *Popular Mechanics*, 60 (2), Aug 1933: 202–5.

Fig. 3.3 Confusion from above: aerial urban view (*London Opinion* 1 (2) (1939): 77; uncredited photographer).

technologies and then to amateur close-up photography—were equally radical. They employed bold new perspectives to deconstruct conventions and to instil new political, economic and perceptual orientations, respectively. Audiences that initially were discomfited were ultimately seduced. The cultural excursions of holograms a half-century later followed the same road, but found it well-trodden.

Fig. 3.4 Disorienting close-up views (Korth, F. G., 'Take it close up', *Popular Mechanics*, 81 (1), Jan 1944: 118–21; illustration p.120).

3.4 IMAGERY TO EMULATE

Popular audiences were active participants in this new visual dialogue, too. Photography attracted and transformed not just the general public, but also enthusiasts and hobbyists—a cultural niche that later contributed to the popularity of holograms. As suggested by the avocation of William Abney (Section 2.2), amateur enthusiasm for photography was evident from its early days, at least among technical proficient and relatively affluent practitioners.[20]

Photography at the turn of the twentieth century had been exclusive, owing to the high cost of cameras. The original Kodak box camera sold for $25, and processing each roll of 100 negatives, developing prints and supplying the reloaded camera cost a further $10, the equivalent of an expensive high-end camera today.[21] By energetically adapting to new audiences, however, Kodak grew over the next century to a near monopoly position in American photography, capturing some 90% of the growing demand. It did so by servicing markets ranging from scientists and engineers to aerial and professional photographers, and from technical amateurs to unskilled consumers. The company oriented its research laboratories and publications to these distinct markets, and its products ranged from a diverse assortment of cameras and photosensitive materials to chemicals and ancillary equipment.[22]

For some consumers, hobby photography proved to be an attractive alternative to Kodak's original integrated socio-technical product. Autonomy and creativity could be fostered by breaking the company's chain of linked activities. By purchasing disparate equipment and chemicals, amateurs could develop their own films and print them in a home darkroom at modest cost. For photographic manufacturers—including Kodak—a new market opened to supply specialized camera accessories, chemical formulations, developing tanks and trays, enlargers, safelights, print dryers and more. Darkroom workers gained the satisfaction of mastering chemical and physical arts as had their Victorian counterparts a generation earlier, but had considerably more latitude in creating high-quality photographs. A community dimension was supported by the growth of photography clubs and hobbyist magazines. By the end of the First World War, darkroom work and hobby photography were the model for a new breed of technical enthusiast, a theme that was to underpin the later appeal of holograms.[23]

[20] On the first wave of (Victorian) amateurs, see Seiberling, G. and C. Bloore, *Amateurs, Photography, and the Mid-Victorian Imagination* (Chicago: University of Chicago Press, 1986).

[21] The camera plus set of developed prints were about $700 and $250, respectively, in 2015 currency. Kodak introduced the Brownie camera in 1900 priced at just $1, but by that date film and processing were available separately from alternative sources.

[22] Jenkins, R. V., *Images and Enterprise: Technology and the American Photographic Industry 1839 to 1925* (Baltimore: Johns Hopkins University Press, 1975); Jenkins, R. V., 'Technology and the market: George Eastman and the origins of mass amateur photography', *Technology and Culture* 16 (1975): 1–19; Munir, K. A. and N. Phillips, 'The birth of the "Kodak Moment": institutional entrepreneurship and the adoption of new technologies', *Organization Studies* 26 (2005): 1665–87.

[23] Discussed more generally in Chapter 8. For the evolving activities of photographic amateurs, compare contemporary texts, e.g. Lund, P., *Photography as a Hobby* (Bradford.: Percy Lurid, Humphries & Co, 1895) versus Davis, W. S., *Practical Amateur Photography* (Ann Arbor, MI: Little, Brown & Co., 1923).

Serious photographic hobbyists, along with the economic and social contexts to support them, were consolidated through the interwar period and Second World War. Their technological enthusiasms developed in part through books and articles written by scientists to popularize science, but at least as much from companies and clubs.[24] Through imaginative advertising, informational publications and constant product development, photographic companies such as Kodak, Ilford and Agfa developed hobbyist markets in North America and Europe.

An important part of the popular appeal, and a template for later hologram enthusiasts and firms, was the technical fertility of hobby photography. The amateur could concentrate, for example, on camera selection and use, benefiting from a stream of varieties, models and specifications having gradually falling prices. The technical advantages of a pre-World War I large-format plate camera (high resolution, subtle gradations of tonality and compensation mechanisms for depth of field and perspective) versus a 'miniature' camera such as the 1925 Leica (portability, speed, film cost) might lead a purchaser to the technical compromise of a 1937 Rolleiflex (large viewfinder for accurate framing, simple film loading and automated exposure counting). The challenges of photographing subjects in difficult lighting, varying distances (close-ups or long shots) and speeds motivated new accessories and perennial discussions—a characteristic inherited by modern technical pastimes, too.

On the other hand, enthusiasts could focus attention on the processing of the image. The optimal choice of film and photographic paper was complicated by a growing range of commercial offerings. The selection and use of the chemicals for development of the negative and processing of the print demanded familiarity with a spectrum of potential ingredients and even theories of chemical image formation. The field offered the seduction of alchemy and the potential of creating new wonders of image quality, consistency or economy. Beginning hobbyists might rely on the formulations recommended by their favoured company or the salesperson at their local photography shop, while advanced hobbyists could peruse technical articles supplied by manufacturers or photographic journals and mix their own. And darkroom equipment, from electric timers to Bakelite developing tanks to optical diffusing systems for enlargers was continually evolving and competing.[25]

The most technically competent hobbyists disseminated their own successes in much the way that Abney had pioneered decades earlier. And at the simplest do-it-yourself level, labour-saving or frugal tricks to build camera accessories or darkroom equipment became popular in magazines between the wars as 'photo kinks', a precursor of twenty-first-century channels of information such as YouTube (Figure 3.5). Through such avenues, visual creativity was democratized.[26]

[24] On the former, top-down route, see Bowler, P. J., *Science for All: The Popularization of Science in Early Twentieth-Century Britain* (Chicago: University of Chicago Press, 2009).

[25] As discussed in Chapter 8, similar processing innovations were also at the heart of holography during the 1970s.

[26] Carlson, S., *Photo-Kinks: The Photographer's Handy Book* (Minneapolis, MN: Huddle, 1937). Such articles were popular in general magazines such as *Popular Mechanics* (founded 1902) and *Everyday Science and Mechanics* (founded 1931).

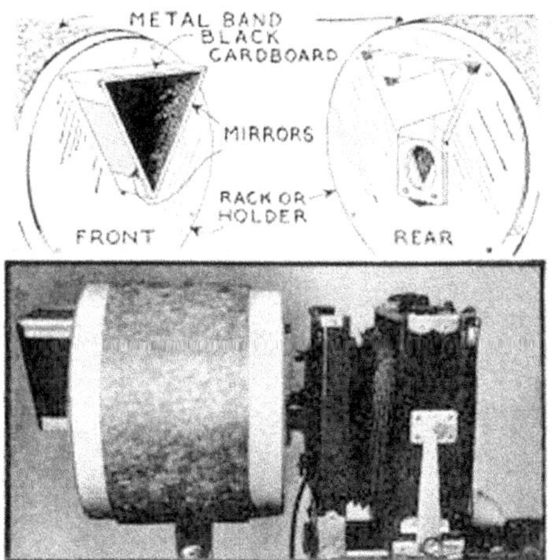

Fig. 3.5 Democratizing spectacle: kaleidoscopic pictures at home ('Unusual photos made with camera mirrors', *Popular Mechanics* 74 (3), Sep 1940: 406; uncredited illustrations).

Like William Abney some decades earlier, the serious photographic amateur straddled the line of amateurism and professionalism. He or she studied the technical data supplied by manufacturers, understood analyses of the type championed by Hurter and Driffield, and often aspired to a systematic experimental approach to their hobby to understand and control the multitude of variables affecting the quality of the photographic image.

Novice or expert, enthusiasts who stuck with photography would find themselves gradually encultured with specialized knowledge having an increasingly scientific cachet. Many could recognize vagaries of image quality such as excessive grain, motion blur, poor focus, low shadow detail or flare; the most adept could attribute these to film selection, shutter speed, lens technique, processing errors or lack of lens hood. Expert amateurs could compensate for some of these faults by retouching negatives, employing texture screens over the paper to be printed, 'dodging' and 'burning' parts of the image projected by the enlarger onto photographic paper, and intensifying or reducing areas of the print by chemical treatments. And the more aesthetically refined hobbyists could crop, tone, vignette or create montages of their prints through additional darkroom techniques. For the ambitious amateur, then, a spectrum of competences from technical to aesthetic was accessible.

Individual expertise was supported by a variety of specialist publications. The primary source was manufacturers' catalogues, information pamphlets and data compendia informed by the research labs operated by many photographic companies.[27] Enthusiasts'

[27] See, for example, Bothamley, C. H., *Ilford Manual of Photography* (London: Brittania Works Co., 1890); Eastman Kodak Co., *Kodakery: A Magazine for Amateur Photographers* (Rochester, NY: Eastman Kodak Co., 1913); Eastman Kodak Co., *Photography: An Outline Course for Instructors* (Rochester, NY: Eastman Kodak Co., 1926); *Kodak Data Book on Copying* (Rochester, NY: Eastman Kodak Co., 1941); *Kodak Master Photoguide* (Rochester, NY: Eastman Kodak Co., 1951).

magazines were a more rapidly expanding resource.[28] And, from the 1940s, specialist books served both scientific professionals and serious hobbyists.[29] In photography and cinematography, amateur skills could approach those of professionals.[30]

More direct social interactions were also important in building the pastime and its skill-set. A handful of societies nurtured a professional connection. The Photographic Society of London, founded in 1853 to promote the 'Art and Science of Photography', for instance, had closely followed on from the influential Great Exhibition of 1851 and the fashion for technological advance recounted in Chapter 2. Its remit was expanded to the whole country in 1874, and became the Royal Photographic Society of Great Britain (RPS) in 1894. Like a professional institution, the RPS was hierarchical in recognizing different levels of technical distinction.

At the other extreme were local camera clubs, providing hobbyists with a regular venue to share their enthusiasms. The availability of small-format but relatively sophisticated cameras from the 1920s encouraged an expansion of hobbyists, and during the 1930s camera clubs flourished. In late 1933 the Photographic Society of America (PSA) was founded as an umbrella organization. In its first year some fifty-one local clubs were established, and by 1938 some 373 were affiliated with the PSA. Most of these clubs, like the RPS, valorized expertise and accomplishment. In addition to meetings and group photographing activities, clubs modelled competitive exhibitions on the art salons of the nineteenth century. The Miniature Camera Club of Philadelphia, for example, instituted the First Salon of the Miniature Camera in 1935 in association with the Philadelphia Art Alliance; two years later it organized the First National Photographic Salon for Women. The vogue for juried photographic competitions communicated the cultural norms of good photography and particularly styles of composition, allowing hobbyists to be ranked, and often rankled.[31]

Participants became versed in technical details that had been little-known or valued at the turn of the century. For the most proficient among them, photographs could now be interpreted and judged according to technical and aesthetic criteria. The cultural significance is that this new refinement extended not merely to artists and the small community of privileged connoisseurs of the previous century, but instead to a swelling community

[28] Seminal English-language hobby photography magazines included *Amateur Photographer* (London: 1884); *Camera Craft* (San Francisco: 1904); *American Photography* (New York: 1907); *Popular Photography* (New York: 1937); *Modern Photography* (New York: 1937); and, *Everyday Photography* (1937).

[29] Popular examples include Jordan, F. I., *Photographic Enlarging* (Rochester, NY: Folmer Graflex Corp, 1935); McKay, H. C., *The Photographic Negative* (Vols. I–IV) (New York: Ziff-Davis, 1942); Lootens, J. G., *On Photographic Enlarging and Print Quality* (New York: The Camera, 1944). Some publishers, notably Focal Press, specialized in publications for amateur photographers, such as Mason, L. C., *All About Making Darkroom Gadgets With Your Own Hands* (London: Focal Press, 1954).

[30] Zimmerman, P. R., 'Professional results with amateur ease: the formation of amateur filmmaking aesthetics 1923–1940', *Film History* 2 (1988): 267–91.

[31] Schwarz, D., 'Camera club photo-competitions: An ethnographic approach to the analysis of a visual event', *Research on Language and Social Interaction* 21 (1987): 251–81. See also Schwartz, D. and M. Griffin, 'Amateur photography: the organizational maintenance of an aesthetic code', in: T. Lindlof (ed.), *Natural Audiences: Qualitative Studies of Media Uses and Effects* (Norwood, NJ: Ablex, 1987), pp 198–224.

of middle-class consumers. The culture of amateur photography was sustained by this constellation of companies, hardware, publications and communities.[32]

The themes of individual innovation and experimentation that had become fashionable for photographic hobbyists during the 1930s were consolidated during the war, when the merits of clever thriftiness were combined with a dearth of new consumer products. By the late 1940s, then, amateur photography was an established pastime having a strong component of scientific and engineering innovation. Both themes were to be important in the cultural popularity of holograms a generation later (explored further in Chapter 7).

3.5 IMAGERY TO ENTERTAIN

As popular engagement with imaging grew, visual technologies continued to innovate. Avant-garde graphics made photographs unsettling, but motion picture technologies sought to attract audiences by new forms visual dissonance. The Lumière brothers' 1896 film of an approaching steam locomotive famously startled audiences unaccustomed to entertainments that threatened to leave the stage.[33] More frequently, though, cinematic innovations provided a startlingly improved approximation of visual reality. Colour processes, exhibited experimentally from the turn of the century, sought to transform the monochromatic image into something more lifelike (although the earliest hand-tinted frames could accentuate the visual surprise merely by showing red fire or unfamiliar below-the-sea views tinted blue).

Further immersion of viewers within the image also generated wonder and appeal. Widescreen formats were trialled from the 1920s, with one scheme touted as a radically new and natural cinematic experience because it more closely mimicked the field of view of two eyes.[34] The most ambitious of these technologies was *Cinerama*, which relied on three synchronized projectors to illuminate a screen that covered a viewing arc of 146 degrees for audiences of up to 800.[35] Its 1952 premiere included scenes from a roller-coaster and helicopter. A New York newspaper reviewer focused on the perceptual shock: the audience reaction was as 'excited and thrilled by the spectacle presented as if it were seeing motion pictures for the first time'.[36]

Motion pictures revived some cultural desires of earlier imaging technologies, just as holograms would refocus them again. For example, there had been a significant overlap between early cinema and the parlour technology of the stereoscope. In 1912, an

[32] 'Stereophotography sweeps the country', *Popular Mechanics*, 98 (3), Sep 1952: 106–11.

[33] Lumière, A. and L. Lumière, 'L'Arrivée d'un train en gare de La Ciotat', (1895). See also Gunning, T., 'An aesthetic of astonishment: the (in)credulous spectator', *Art and Text* 34 (1989): 31–45.

[34] 'Now comes the natural vision movie – which looks like the pictures you used to see thru the stereoscope', *Motion Picture Magazine*, 33 (1), Feb 1927: 57.

[35] Even more ambitious multi-projector panoramic film configurations had been created for special projects, notably an 11-projector system, 'Vitarama', for the 1939 World's Fair in New York City.

[36] Crowther, B., 'This is Cinerama: new movie projection shown here; giant wide angle screen utilized; novel technique in films unveiled', *New York Times*, 1 Oct 1952, 1 40.

executive of Selig Polyscope Company, an early motion picture firm, mused about the future for cinema. Instead of being relegated to boardwalk entertainments and makeshift viewing rooms, he forecast, the motion pictures of the future would have colour, sound, immediacy and depth; it would be a fully *natural* reproduction of reality amenable to personal use, in which the picture would 'stand out from the screen in a bold, natural relief, instead of being flat'.[37]

Stereoscopic films had been created and exhibited experimentally since the beginning of cinema; indeed, the Lumière brothers had combined all forms of visual imagery in a *tour de force* unmatched again for decades: their famous approaching train had been filmed with two 35 mm cameras, and shown to some audiences via two-colour (anaglyphic) stereo glasses. For postwar Hollywood, however, 3-D movies promised further spectacle and competitive distancing from increasingly familiar home television broadcasts. In the same year that Cinerama was debuted, an unexpectedly popular adventure film, *Bwana Devil*, launched a craze in American movie theatres for stereoscopic movies. In 1953, over thirty 3-D movies, employing more than a dozen commercial processes, were shown on American screens, and similar examples that year were produced in Italy, Mexico, the USSR and Japan.

3-D movies consolidated this popular interest in three-dimensional entertainment, and made stereography for adults briefly fashionable again (Figure 3.6). Promoters' hype could claim that stereoscopic cinema was a natural human attraction, an inevitable technology and even a promoter of health. One 1953 paper published at the height of the stereo boom outlined 'therapeutic benefits from viewing stereoscopic motion pictures' and the 'potent stimulation to good binocular vision'—rehearsing the earlier themes of quack medicine shows.[38]

The cinematic processes, alternately dubbed 'Tri-Dee', 'Future Dimension', 'Tru-Stereo' or 'Natural Vision' by their various adopters, shaped the content to the technology and its potential for shock. Where widescreen films sought to awe viewers with sublime airborne travelogues and 'pictorial spectacles—without any attempt at storytelling', as a *New York Times* reviewer put it, 3-D movies tried instead to surprise and jolt their audiences. The shock relied on the moving objects that seemed to travel right up to the viewers' seats.[39] Many mirrored *Bwana Devil*'s leaping lions and near-tactile romantic interludes: *House of Wax* contrasted the thrill of a three-dimensional fight with a comically deep table tennis game; and *Dial M for Murder* had audiences avoiding a lunging hand. Each sought to discomfit and engage their viewers in ways that wide-screen motion pictures and stereoscopes could not.[40]

[37] Twist, S. H., 'The picture of the future: a reverie', *New York Clipper*, 17 Aug 1912, 7.

[38] Sherman, R. A., 'Benefits to vision through stereoscopic films', *Journal of the Society of Motion Picture and Television Engineers* 61 (1953): 295–308.

[39] For a contemporary technical survey, see Zone, R., *Stereoscopic Cinema and the Origins of 3-D Film, 1838–1952* (Lexington, KY: University Press of Kentucky, 2007).

[40] *Bwana Devil* (Dir. Arch Oboler, United Artists, 1952); *House of Wax* (Dir. André de Toth, Warner Bros, 1953); *Dial M for Murder* (Dir. Alfred Hitchcock, Warner Bros, 1954).

THREE-DIMENSIONAL lobby display piece designed to promote "House of Wax."

Fig. 3.6 Selling 3-D (Blumenstock, M., 'The House of Wax campaign', in: M. Quigley Jr (ed.), *New Screen Techniques* (New York: Quigley Publishing, 1953), pp 93–101, illustration p.99. Courtesy of W. J. Quigley).

The division between the viewer and the viewed was a familiar convention in early twentieth-century theatre, and explains the surprise of audiences viewing the Lumières' train. But this invisible boundary—the 'fourth wall' familiar to actors—remained potent for cinema-goers (and later viewers of 3-D TV), and previewed one of the perennially unsettling attractions of holograms: the viewer's ability to interact with an image that may extend beyond the boundary of the hologram itself.[41]

The fad for 3-D movies fluctuated as audiences first experienced and then rapidly acclimatized to the visual spectacle. During the 1950s, the world-wide peak for productions was thirty-nine (1953), falling to eighteen in 1954 and merely a couple each in 1955 and 1956. Subsequent booms about a generation apart (and thus capturing fresh audiences) were more modest.[42]

3.6 DISPLAYING PROGRESS

The postwar innovations in motion pictures focused increasingly on technologies not only to deliver spectacular images but also as themes of entertainment. Several factors were responsible: the now-established twentieth-century focus that linked modernity with technological progress; the sacrifices of the war, which had both delayed the delivery of new consumer technologies and heightened anticipations of rewards to come; and, public revelations of the genuine technical innovations that had been produced by wartime development.

While living through a decade of unprecedented economic depression, pre-war audiences had come to see technological progress as obvious and inexorable. Popular magazines had chronicled the rapid development of aviation and automobiles, electrical inventions and ambitious architecture. Each of these technologies heralded a better future through the convenience, leisure and entertainment they would provide to a growing population of middle-class consumers, and those aspiring to join them.[43]

Technologists and scientists were receptive audiences, too. Technical advertising from the 1930s increasingly had adopted a rhetoric documenting the steady advance of products through the application of science. Large companies, particularly those lubricated by government contracts for military supply, employed series of advertisements to illustrate the growing diversity and reliability of their technical advances, and often the scientific methods responsible for them. *Scientific American* magazine became an important conduit for such messages: advertisements preached themes of corporate ingenuity, progress and rational science to accompany dispassionate popularizations of the

[41] The fourth wall is often bypassed in children's theatre, where actors may speak directly to the audience. This 'childish' or playful dimension parallels the unfamiliar perspectives of Rodchenko's graphic art, and arguably is a factor in the appeal of holograms.

[42] Sammons, E., *The World of 3-D Movies* (Valencia: Delphi, 1992).

[43] For an overview, see Stearns, P. N., *Consumerism in World History* (New York: Routledge, 2001).

current state-of-the-art in scientific knowledge.[44] More surprisingly, science and technology magazines underlined these themes visually by adopting the disorienting graphic arts that had first appeared in politicized news magazines of the 1920s. Examples were frequent, for example, in the Massachusetts Institute of Technology magazine *Technology Review* and the British magazine *Discovery* (Figure 3.7). The imagery rehearsed the

Fig. 3.7 Visual surprise and modernity (*Discovery: The Magazine of Scientific Progress*, 8 (11), Nov 1947: 1; uncredited photographer at Bell Laboratories).

[44] Companies such as Bell Labs, General Electric and Du Pont regularly provided full-page advertisements. As discussed in Sections 7.2 and 7.3, *Scientific American* also became a major channel for amateur science by promoting the spirit of individual enthusiasm allied with hands-on technical competence.

pre-war graphic arts and the perspective-play introduced by current-events magazines, but realigned it more enduringly to the progress of science and technology. Illustrations of high-technology visual sciences also frequently featured, with stories on microscopy, stroboscopes and stereoscopy in science. Stereo imaging was fashionable enough to serve as a metaphor for the editorialists' aims:

> The *Technology Review* is designed to be an editorial stereoscope for presenting Science's new world picture in the startling clarity of relief. It goes beyond the mere reporting of science and engineering news in plano; it adds the third dimension of *interpretation.*

It continued:

> It is for such readers that The Review presents each month a salon of brilliant photographs with an able supporting cast of three-dimension captions. In these illustrations, readers find both interpretation and beauty, the unexpected, otherwise unseen beauty caught by camera-in-the-laboratory and by engineer-inspired cameras on the mammoth dams and bridges and in the great industries of today and tomorrow.[45]

For diverse audiences, stark graphic arts outlined intangible cultural convictions: the modern postwar world would be simultaneously harsh, thrilling and progressive. Advancement was inexorable, and new technologies would sweep us into the future.

The rhetoric of progress was equally seductive to postwar audiences, even where rationing still limited the satisfaction of consumer aspirations. For North Americans, arguably among the best provisioned by resources, government and manufacturers had preached wartime frugality and patience. By 1944, with a successful end to the war in sight, companies prepared audiences for a return to full civilian production. Consumers anticipated not merely resuming consumption at pre-war levels, but catching up for lost time. In Britain, the end of the war also signalled expectations of a levelling of class differences, and a consequent improvement in the standard of living of the lower-income majority. The availability of material luxuries would be combined with a new freedom to enjoy them. And in the many other countries ravaged or occupied during the conflicts, the postwar exposure to western products often created new aspirations. An important feature of the postwar world was that these desires were manufactured by governments as well as industries. The Marshall Plan for European recovery (1948–1952), for example, was motivated by the American government's aim of preventing the expansion of communism in Western Europe. Its aid included essential monetary support and food, but also specified conditions that would provide export markets for American products. In France, this included the obligation to screen dubbed American films; in Britain, petroleum refining and automobile manufacture expanded under American companies.

[45] Rowlands, J. J. and J. R. Killian Jr, 'Seeing solid: the third dimension at work and play', *Technology Review*, 39 (5), Mar 1937: 191–5; editorial p. 182; italics in original. The transition from paintings and drawings to almost exclusively modernist photographic illustrations began for the periodical with a few corporate advertisers in mid-1931, followed by contents page photos that autumn and finally dramatic cover photos from November 1931. Margaret Bourke-White was a long-time contributor, as were practising MIT engineers; older photos were also cropped for dramatic effect by the editors.

During the same period, American security agencies identified modern art and music as subtle expressions of American freedom and innovation. Thus art and technology became more explicit cultural exports.[46]

Postwar expectations of consumer goods focused on labour-saving appliances and luxuries. Some were improved models of familiar devices such as stoves, cars and radios. Others were products that were not yet part of average homes, like telephones, refrigerators and automatic washing machines. And a few had been widely forecast but experienced by only the most privileged pre-war urban audiences. The most important of them for building consumer culture was television.

Unlike cinema and still-photography, television was not a technological surprise; it had been long awaited. Popular dreams of electric imaging had closely followed the invention of the telephone in 1878.[47] Practical television evolved in close association with broadcast radio from the early 1920s, and its rumoured characteristics—usually far in advance of technical achievements—circulated through the 1930s. Demonstration systems for urban audiences were implemented briefly in the UK, Germany and America during the 1930s, but mothballed during the war.

Expectations of television also rehearsed recent cultural experiences with other media, and provided a preview of how holograms would later be received. Three-dimensionality was an enduring aspiration (Figure 3.8). Indeed, the Scots inventor, John Logie Baird, had contrived a crude demonstration of stereoscopic imagery using his system of mechanically-scanned television. He continued to innovate with electronically-scanned systems from the mid-1930s, leading to demonstrations for the press and British government representatives during the early 1940s of colour and stereoscopic TV based on two-colour (anaglyphic) images. Critical assessments mirrored earlier technologies and holograms to come. A contemporary BBC report summarized that

> its effectiveness appeared to vary with each individual's sight. It seemed to create a greater sense of depth between the centre of the picture and the background. The greatest objection to it is that one has to wear peculiar glasses in which one side is coloured blue and the other red.

The British Admiralty responded politely by suggesting to Baird that such systems might find peace-time applications.[48]

As with stereoscopes and radio, an educational role was forecast for television to complement its appeal as mere mass entertainment. Hobbyists identified new attractions to

[46] Hogan, M. J., *The Marshall Plan: America, Britain, and the Reconstruction of Western Europe, 1947–1952* (Cambridge: Cambridge University Press, 1987); Saunders, F. S., *The Cultural Cold War: The CIA and the World of Arts and Letters* (New York: W. W. Norton, 1999).

[47] *Punch's Almanac* for 1879 facetiously reported that 'Edison's telephonoscope', an 'electric camera obscura', would make possible international conversations via a wide-screen and high-resolution picture (du Maurier, G., 'Cartoon', *Punch's Almanac*, 8 Dec 1879).

[48] See Kamm, A. and M. Baird, *John Logie Baird: A Life* (Edinburgh: National Museums of Scotland, 2002), pp 338–40, and Burns, R., *John Logie Baird, Television Pioneer* (Stevenage: Institution of Electrical Engineers, 2000), pp 367–72; Schuster, L. F., 'Memorandum to the Director General (BBC)', British Broadcasting Corporation, 28 Apr 1942.

Television in Three Dimensions

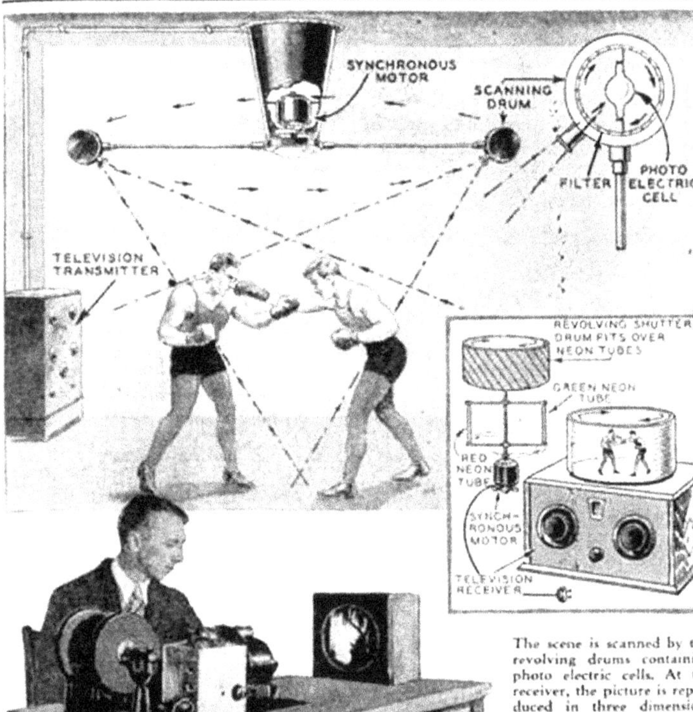

The scene is scanned by two revolving drums containing photo electric cells. At the receiver, the picture is reproduced in three dimensions with neon tubes, as shown.

The scene can also be reproduced on a screen by means of a television controlled projector.

A DEVICE which can produce a 360 degree picture by television through a stereoscope scanner has been invented by Leslie Gould, a radio engineer of Bridgeport, Connecticut. With Mr. Gould's television system it is possible to televise a boxing match, a play, an orchestra, or any other spectacle whose scene of action can be compressed into a reasonable space.

The new invention makes use of neon tubes of various sizes and colors, depending upon the magnitude of the image. The spot on which the television subject is located is scanned by beams from two rotating arms, as shown in the drawing above. At the extremity of each arm is a scanning drum containing a photo electric cell, which picks up the images to be televised.

The scene which is going to be televised must be flooded with a great quantity of strong light. Each electric eye catches the light reflected through the apertures in the revolving drums from the light portions of the body, and flashes to the transmitter the electric impulses set up by the variations in light.

Inventions for February 67

Fig. 3.8 Imaging in the round ('Television in three dimensions', *Modern Mechanics and Invention*, Feb 1931: 67, uncredited illustrator. Charles Shopsin collection).

revitalize radio expertise and impress friends and family; technicians and advertisers were seduced by new opportunities for founding and promoting small businesses.

Postwar consumer expectations of television were consequently high. Magazines that had been dedicated to radio over the two previous decades now previewed, reviewed and forecast television programmes and receivers. Picture size and quality increased steadily.[49] Both television and photojournalism highlighted actuality and immediacy, and sought to impress viewers with the pace of change. Conceived as a medium of entertainment and education before the war, television became a conduit for understanding modern science after the war.[50]

The end of the war itself was marked by revelations of secret inventions and their powers. Schemes for the automatic aiming of artillery—and the scientific potential of automation and computation—captivated readers. The invention of the transistor in 1947 at Bell Laboratories, validated by the Nobel Prize for Physics nine years later, was touted as a triumph of science that would allow not only compact and rugged military gear, but also low-powered portable consumer products.[51] Its final year had been accompanied by news of vengeance weapons developed by Nazi Germany: the V-1 cruise missile or 'buzz bomb' first used in the week after the D-Day invasion in June 1944, and the V-2 ballistic missile from September. The Allied war against Japan was ended dramatically the following August by atomic bombs of two different designs. Even more significantly for shaping civilian attitudes, the details of these technologies remained under wraps. V-2 rockets, and entire German research teams working in a variety of fields, were captured by the Americans, British and Soviets for further debriefing and development; the principles of nuclear weapons, and the vast wartime American factories for their production, were briefly sketched in postwar publications, but then just as quickly hidden.[52]

These revelations implied further unsuspected technological innovations and the potential for more. Publics habituated to scanning the skies for enemy aircraft during the war proved to be receptive audiences for the first reports in June 1947 of unidentified flying objects having physics-stretching capabilities. Even the growing mythology of UFOs and suspicions of alien presence fitted mentalities cultivated during the war: such disturbing wonders naturally would be kept under wraps by governments seeking to protect their citizens and to gain a technological advantage over their enemies. Scientific advance, technological powers and secrecy made natural bedfellows.

The central role of science and technology in the outcome of the war and in the postwar world was appreciated by all (Figure 3.9). For governments as well as their publics, the enduring lesson was that investment in science could guarantee the solution of seemingly insoluble problems.[53] Government programmes were instituted to fund and direct science towards nationally-important problems by every country that could afford it.

[49] Moran, J., *Armchair Nation: An intimate history of Britain in front of the TV* (London: Profile, 2013).

[50] Lafollette, M. C., *Science on American Television: A History* (Chicago: University of Chicago Press, 2013). On visual portrayals of science, see Ford, B. J., *Images of Science* (New York: Oxford University Press, 1993).

[51] The first public account of the transistor reported that 'The device was demonstrated in a radio receiver . . . in a telephone system and a television unit' ('The news of radio', *New York Times*, 1 Jul 1948: 46).

[52] Johnston, S. F., *The Neutron's Children: Nuclear Engineers and the Shaping of Identity* (Oxford: Oxford University Press, 2012), Chapter 4.

[53] del Sesto, S. L., 'Wasn't the future of nuclear energy wonderful?', in: J. J. Corn (ed.), *Imagining Tomorrow: History, Technology, and the American Future* (Cambridge, MA: MIT Press, 1986), pp 58–76.

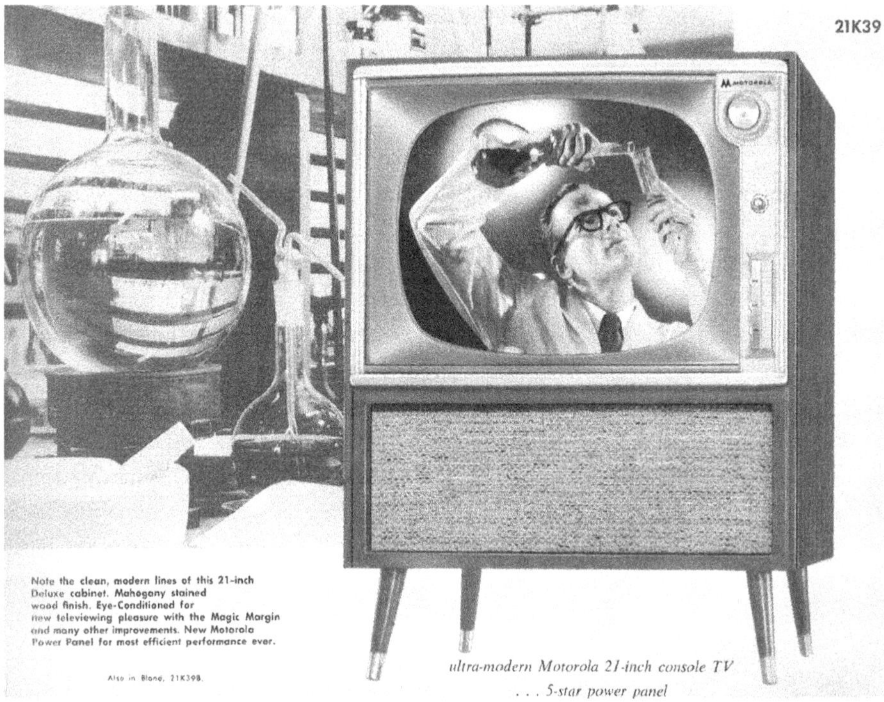

Fig. 3.9 Advertising television receivers and science (Motorola (1956). S. Johnston collection).

Big Science, and especially its application to military development, became the dominant approach for the remainder of the century.[54]

Modern culture absorbed and embraced this new vision of science through its technological spin-offs. Consumers increasingly recognized themselves as the beneficiaries of a chain of industrial processes fostered by science. Just as wartime radar and aircraft design were now yielding safer and faster civilian aircraft, consumer products could advertise the scientific discoveries behind their development.

The immediate prelude to holograms, then, was a postwar culture that had been prepared for rapid progress in scientific knowledge and consumer technologies, and a visually literate public sensitized to the quality and impact of imaging. Ironically, then, the research that led to holograms took place not in the glare of popular culture but in secretive environments that had been fostered by the war.

[54] See, for example, de Solla Price, D., *Little Science, Big Science* (Washington: Columbia University Press, 1963); Weinberg, A. M., *Reflections on Big Science* (Boston: MIT Press, 1967); Galison, P. and B. Hevley (eds.), *Big Science: The Growth of Large-Scale Research* (Stanford: Stanford University Press, 1992).

PART B
Making Sense of Holograms

Holograms were conceived in postwar laboratories as a key element of sophisticated new microscopes and radar systems. For over a decade after their invention, they were seen by no more than a few dozen people, and inspired even fewer.

When combined with newly-invented lasers, however, holograms revealed magical qualities. Their startling properties challenged description, and the revelation of holograms to wider audiences was a celebration of the modern world.

4

Hologram Secrets

Holograms were not among the shock troops of imaging until two decades after the Second World War. Instead, they inhabited relatively hidden niches of technology. Holograms were invented and promoted at the frontier of electronics and microscopy during the late 1940s, a period when science and technology were in catch-up mode but weighed down with the burden of wartime associations. Frustrated by the recent sacrifices but exhilarated by the revelation of secret developments, the public and technologists alike were at a cusp of confidence concerning scientific advance.

Founded on the unintuitive and complex mathematical understanding of imaging that had begun a half-century earlier, holograms epitomized this faith in progress. The first small audiences for the arcane technique were microscopists and scientific instrument companies.

These postwar cultural dimensions were neither popular nor even *un*popular. They remained largely unknown to wider publics, and survive sparsely represented in archival records and oral histories. The contexts in which the first holograms were explored involved furtive but intensive investigations in different corners of the world by professional scientists and engineers. Their Cold War activities sought to develop technologies in the national interest. Scientific cultures, then, nurtured the promising new principles of imaging; wider cultures discovered holograms later.

This chapter explores a hidden domain that was the antithesis of popular culture: the earliest exploration of holograms as tools of engineering during a period of rising secrecy. Holograms required a technological side-step and a cultural *pas de deux*. The subculture of scientific photography—the narrative focus of Chapter 2—was partnered with the distinct perspectives of electronic imaging. This account traces those activities to highlight how hybrid technical cultures developed and how they motivated research and applications of imaging. The shadowy work defined the characteristics of the first holograms and how they were portrayed to early audiences.[1]

[1] For an extensive historical account, see Johnston, S. F., *Holographic Visions: A History of New Science* (Oxford: Oxford University Press, 2006), Chapters 2–5.

4.1 STARTING SMALL: MICROSCOPIC CULTURE

The experiments with the optical model have progressed very satisfactorily and I have produced reconstructions much better than those in the report, and in the short article which has recently appeared in Nature. I wonder if it is not yet too late to suggest a talk on this subject at the forthcoming meeting of the British Association?... I should be very glad to give a short talk, and perhaps also to demonstrate the optical model in operation, as I have done at the April conference of the Electron Microscope Group, where it had a very satisfactory reception ... [2]

Dennis Gabor, Rugby, England, 1948

The *hologram*, from Greek roots denoting *whole picture*, was conceived and christened in 1947 by Dennis Gabor (1900–1979), a Hungarian engineer-physicist working in England. Gabor had left Hitler's Germany in 1933 for an electrical engineering post at the British Thomson-Houston Company (BTH) in Rugby, England. Lacking security clearance during the war, he had worked on projects at BTH that did not have military significance. These low-priority and long-term studies allowed free rein for Gabor's creativity. Three lean years after the war allowed him to develop these ideas further.

Gabor was working in an electrical engineering environment, but one that was expanding towards electronic imaging. His prior experience as a PhD student in Germany was relevant. Gabor had designed a cathode-ray oscillograph, a device to record rapid electrical signals. Its central feature was a beam of electrons that could be focused and directed by an applied magnetic field. His apparatus was subsequently scavenged by later students to create the first electron microscope. Ernst Ruska (1908–88) reconceived the heart of Gabor's apparatus as an 'electron lens' and combined such elements to yield an electronic device that could magnify images much like an optical microscope did. In the electron microscope, an electron beam was scattered off the physical sample and then focused with magnetic lenses to create an image on a fluorescent screen. Ruska was to gain the Nobel Prize in Physics in 1989 for this extension of microscopy via electronics.[3]

Gabor followed the development of electron microscopy in Germany and other countries. His lost opportunity was made more annoying by the parallel commercial development of electron microscopes at a British sister-company, Metropolitan-Vickers. Consulting for Met-Vick from 1943 allowed him to imagine developing electron microscopy further. Indeed, he wrote papers on electron lenses during the war and his first book, *The Electron Microscope*, in 1948. Electron optics became one of his professional skills before the term was invented.[4]

[2] Gabor, D. to L. Bragg, letter, 6 Jun 1948, Rugby, IC GABOR EL/1.

[3] Kunkle, G. C., 'Technology in the seamless web: "success" and "failure" in the history of the electron microscope', *Technology and Culture* 36 (1995): 80–103.

[4] Gabor, D., *The Electron Microscope: Its Development, Present Performance, and Future Possibilities* (London: Electronic Engineering, 1948). This research on the steering and containment of electron beams was seminal for his career. During the 1950s as Reader in Electronics at Imperial College, London, Gabor's principal research projects involved investigating the design of television tubes and, later, the containment of charged particles for nuclear fusion.

But Gabor also devoted time to more conventional optics, too, and specifically stereoscopic imaging. In 1939, BTH engineers had listened to a speech by Oscar Deutsch, the creator of the Odeon cinema chain, about the future of commercial film. Deutsch predicted that motion pictures would have to be three-dimensional, and called on British engineers to develop a scheme that did not require special viewing glasses. The BTH Research Director encouraged Gabor to work on the problem, and by 1940 he had authored a series of patents for a stereoscopic viewing system suitable for small screens.[5]

Gabor shared some of the attributes of Gabriel Lippmann a half-century earlier. His breadth of experience across electronics, optics and mathematical physics gave Gabor an unusual intellectual perspective, and also visibility among his peers. In his postwar years at BTH, he consequently gained opportunities to do research on more abstract topics that promised longer-term commercial benefits.

The most significant of these concepts was the hologram. Gabor coined the term in 1947 to label a new concept for recording images of a more complete form. His focus, though, was narrow: he envisaged the hologram as an optical component for improving the first generation of electron microscopes.

Gabor's career experience straddling the cultures of electronics, mathematics and optics was crucial to conceiving his concept. He imagined the hologram as an intermediary step for producing high-resolution electron microscope images. His familiarity with mathematical optics and quantum mechanics indicated that electrons could be considered as both particles and waves, and so could be understood as analogous to visible light.[6] Because electronic imaging and optical imaging were closely related, Gabor reasoned that he could merge the technologies: the electron microscope could be improved by adding an optical portion.

Electron lenses could produce extreme magnifications, but they were, in the terminology of optics, 'simple'. Like a single glass lens, they created serious aberrations that limited the image quality. Glass lenses, though, could be constructed from different types of glass and combined to correct such aberrations. The result was exquisitely precise images, but there were still limits. Optical microscopes could only magnify by a factor of a few hundred because they were ultimately limited by the wavelength of light. Any further magnification was swamped not just by exaggerated aberrations, but more fundamentally by the bending of light waves around edges. As Ernst Abbe had reasoned at

[5] Gabor, D. to T. Raison, letter, 15 Jul 1961, IC GABOR MN/5–6; Tanner, P. G. and T. E. Allibone, 'The patent literature of Nobel laureate Dennis Gabor (1900–1979)', *Notes and Records of the Royal Society of London* 51 (1997): 105–20. For an overview of Gabor's career, see Allibone, T. E., 'Dennis Gabor 1900–1979', *Biographical Memoirs of Fellows of the Royal Society* 26 (1980): 107–40.

[6] As a student in Germany during the late 1920s, Gabor had attended lectures by the first generation of modern physicists developing quantum mechanics. He later collaborated with Max Born, recipient of a Nobel Prize, providing a chapter for a mathematically rigorous textbook (Born, M. and E. Wolf, *Principles of Optics: Electromagnetic Theory of Propagation, Interference, and Diffraction of Light* (London: Pergamon Press, 1959)).

the turn of the century, light would be deviated by each aperture, smearing the focused image and obscuring detail.[7]

Gabor's concept involved a two-step imaging process. It would begin with an electron beam as used in electron microscopes, but without the distorting magnetic lenses. First, the beam of coherent radiation (that is, the electron beam equivalent to waves having a single frequency and moving in phase) would generate the shadow of the microscopic object.[8] That shadow, according to physical optics, would be surrounded by fringes of optical interference, because the electrons themselves, considered as waves having a very short wavelength, would interfere constructively and destructively to yield light and dark regions. This process of wave interference and fringe formation had been dubbed *diffraction* by nineteenth-century optical scientists such as Abbe. The wavefront passing very close to the object would be diffracted, or deviated, towards portions of the wave that are undeviated further away. In this way, interference fringes would ring the shadow of an opaque object.

The first step of the imaging process was to record this interference pattern of the electron beam. In Gabor's scheme, this diffraction pattern would be recorded on photographic film and developed.

In the second step, this interference photograph, or *hologram*, would be used to diffract visible light to 'reconstruct' an image of the original subject (Box 4.1). The advantage of this circular procedure is that the reconstructed image would be magnified. The magnification would no longer rely on inadequate magnetic lenses; instead, it would result from the intrinsic connection between electrons and light. Both could be understood as waves, but with grossly different wavelengths; the magnification depended on their ratio. So, this second step would generate a magnified image by lighting the hologram with monochromatic (single-wavelength) light, and viewing the magnified image through an eyepiece—an apparatus that Gabor dubbed a 'holoscope'.

Gabor's revised notion of an electron microscope was thus a curious hybrid. He would abandon its core idea—the magnetic lens—and replace it with a system that would have been familiar to scientists sixty years earlier: a darkroom for processing the hologram and an optical bench and optical system to view the image. The complex and retrograde technique also demanded hybrid users: technologists skilled in both high-voltage electronics and photographic knowhow.

[7] Abbe, E., 'A contribution to the theory of the microscope and the nature of microscopic vision', *Proceedings of the Bristol Naturalists' Society* 1 (1874): 200–61. Abbe's theory of the microscope was a tour-de-force of mathematical optics that allowed him to improve the design of the products of Zeiss, the company for which he worked. The concepts remained unfamiliar for many scientists and most technologists in the mid-twentieth century, however. Gabor's awareness of these fundamental limitations of the microscope, combined with intervening developments in quantum mechanics, helped him to appreciate some of the possibilities for electron microscopes.

[8] Gabor intended eventually to use an electron beam, which has the same wave-like properties as visible light, in the first step of this imaging process, but did most of his experimental work using visible light.

Box 4.1 Viewing Gabor holograms

Gabor holograms (later also known as 'in-line' transmission holograms) are usually fuzzy-looking shadows of a microscopic object ringed by fringes, and are recorded on a small photographic plates a centimetre or two in diameter (See Figure 8.5 for a typical example).

When lit by a filtered (monochromatic) source and viewed through a magnifying eyepiece, a Gabor hologram will focus a sharp image of the object as if it was at the focus of the microscope.

The image quality is low, however: the sharp image is overlapped by another, out-of-focus image and by the glare of the light source itself.

4.2 TRUSTING HOLOGRAMS

Gabor was moving beyond his familiar domain—industrial and theoretical electronics—into new territory inhabited by potential beneficiaries. He found himself in a trading zone, an anthropological context employed by historian of science Peter Galison to describe how different breeds of scientists interact. Galison argued that, in such situations, a shared language, or creole, is constructed to allow the participants to translate and share ideas that are meaningful to each.[9]

In Gabor's case, this meeting of technical cultures proved unprofitable. Communicating his concepts to new audiences, and even to scientific peers, proved difficult. Gabor promoted his technique—relabelled successively *holoscopy, diffraction microscopy* and *wavefront reconstruction*—to varied audiences. His first moves, in the autumn of 1947, were to patent the concept and to speak to a microscopist acquaintance who could publicize it further within his community.[10]

Gabor also lobbied prominent physicists among his contacts with some success. He was informally mentored and guided by two Nobel Prize winners, the British Sir Lawrence Bragg and expatriate German Max Born, who helped edit his publications and to shepherd them towards appropriate journals. He also described his concept at the most public of scientific conferences, the British Association for the Advancement of Science, where scientists, enthusiasts and the media mingle. Gabor sought to explain his ideas in language familiar to mathematicians, optical physicists, practising microscopists and technically-literate laypersons. And, bypassing jargon, he tried to demonstrate how holograms worked to those who stopped by at his conference tables or lab to listen.

[9] Galison, P., *Image and Logic: A Material Culture of Microphysics* (Chicago: University of Chicago Press, 1997).

[10] Gabor, D., Pat. No. 685,286 'Improvements in and relating to Microscopy' (1947), assigned to British Thomson-Houston; Le Poole, J. B. to D. Gabor, letter, 21 Jan 1948, IC GABOR EL/1.

The result was brief attention from wider audiences and *New York Times* coverage that included first mention of the word *hologram*, describing it as looking like a 'futuristic tapestry' having 'no similarity to the object under examination'. In publications for his peers, Gabor emphasized subtle but exciting possibilities. Recording a hologram with an electron beam might allow short exposures, he claimed: contemporary electron microscopes destroyed their specimens as they were examined in real-time, but Gabor's scheme could allow brief recording and then unhurried examination of its proxy, the hologram, by optical means.

But, as earlier promoters of cinema and television had done, and subsequent promoters of holograms would repeat, Gabor attracted wider audiences by combining magical demonstrations with confident technological predictions. The *Times* article focused on the potential of this novel microscopy by focusing on the future:

> By combining the electron microscope with a new optical principle it could be made to resolve the pattern of atoms in the molecule or the details of a virus . . . Moreover three-dimensional objects may be recorded in one photograph.[11]

The promise of recording depth could transform microscopy. Gabor assured audiences that the hologram would allow viewing different depths of the extremely magnified image at leisure:

> It is a striking property of these diagrams that they constitute records of three-dimensional as well as of plane objects. One plane after another of extended objects can be observed in the microscope, just as if the object were really in position.[12]

But Gabor was unable to demonstrate the depth of his technique. Instead, he presented to his audiences a kind of parlour trick: after being shown the fuzzy coin-sized hologram, the viewer would peer through the eyepiece to see a sharper image magically appear.

The demonstration echoed Gabriel Lippmann's colour photography half a century earlier. Most commentators had described the invention second-hand; those privileged to see them seldom had the opportunity to scrutinize the properties of Lippmann's plates closely. Gabor's audiences were similarly voiceless; for the few who were able to witness the effects directly, the brief attraction of holograms was this scientific sleight of hand. Even his confidantes were baffled. Bragg struggled to clarify the mathematics through several drafts of Gabor's articles, and four years later Born admitted that the principle still seemed 'a little weird'.[13]

[11] 'New microscope limns molecule: Britons impressed by paper combining optical principle with electron method', *New York Times*, 15 Sep 1948, 35; Gabor, D., Pat. No. 685,286 'Improvements in and relating to Microscopy' (1947), assigned to British Thomson-Houston. The *Times* story does not appear to have been picked up by any other periodical.

[12] Gabor, D., 'A new microscopic principle', *Nature* 161 (1948): 777–8.

[13] Born, M. to D. Gabor, letter, 21 Feb 1951, IC GABOR MB/10/3. Producing 'deep' images required lasers, which were not available for another fifteen years.

Gabor's new technique came along at an awkward time. The working culture of microscopists was still unsettled by electron microscopes, which demanded non-traditional technical skills. The familiarity of optical microscopy developed over some three hundred years was threatened. Optical microscopists—ranging from biologists to geologists—had a rich portfolio of techniques at their disposal, but little of the intellectual background to appreciate Gabor's concept. The art that electron microscopists had been busily developing over the past decade was also interrupted in full flow. Gabor's short article for the International Commission of Optics generated no response. And while Gabor reported 'a very satisfactory reception' to a poster paper at the London Conference of the Electron Microscope Group of the Institute of Physics in the spring of 1948, there are no contemporary published reports in the specialist literature or surviving archives to back up his claim.[14]

Specialist audiences were bemused at best. Interest was unsustained not only because the concepts were obscure, but because there was a lack of demonstrable utility. Gabor's scheme was plagued by practical and theoretical limitations.

Gabor nevertheless consolidated the idea by using its prospective applications to negotiate a senior academic post at Imperial College, London. Promoting potential export markets for British electron microscopes, he also gained a government research grant to continue research on diffraction microscopy with industrial colleagues. His collaborators over the next decade were unable to stabilize their electron apparatus long enough to record fringes on film.

Through the mid-1950s, Gabor won a handful of academic followers, notably at Stanford University in California, where physicist Paul Kirkpatrick encouraged two students to reproduce and extend his work. Their aim was to use the concepts for x-ray microscopy but, as Gabor and his peers were concluding, experimental limitations looked hopeless. Although he and a handful of others explored the possibilities creatively and tenaciously, Gabor and his audiences had given up the idea of the hologram as a failed invention within a decade.[15]

Gabor's notion was difficult for contemporary audiences to evaluate. The concept was intellectually powerful but far from intuitive; his analysis was mathematically sophisticated but abstruse and inelegant; and, the technology appeared full of potential but also seemed backward-looking and clumsy. Microscopists—the intended beneficiaries—were unconvinced by his efforts.

Holograms began, then, as an awkward technology that held recognizable potential, but that challenged three distinct and burgeoning cultures: those of microscopy, optics and electronics. For each audience, holograms seemed a poor fit to contemporary notions of progress.

[14] Gabor, D. to L. Bragg, letter, 6 Jun 1948, Rugby, IC GABOR EL/1.

[15] Johnston, S. F., 'From white elephant to Nobel Prize: Dennis Gabor's wavefront reconstruction', *Historical Studies in the Physical and Biological Sciences* 36 (2005): 35–70.

4.3 MAKING IMAGERY ELECTRONIC

In the immediate postwar years, science was being realigned. No longer was it an academic or industrial activity alone; it was now both, and carried the new benefits and burdens of increasing government oversight.

Imaging sciences—even established fields like microscopy and astronomy—were now subject to different forces. As suggested by Gabor's difficulties, optics was being merged with a much younger and more dynamic discipline, electronics. Classical optics, with its familiar but staid applications in cameras, binoculars and microscopes, was being extended to invisible radiation (infrared, radar and x-rays); it was being mathematized in unfamiliar ways, such as Gabor's arcane theory; it was being viewed with an eye to novel applications by new sponsors.

With the Cold War in full swing by the early 1950s, research groups increasingly were funded by governments seeking technological developments that would confer military advantage. Aided by being classified, this work shaped distinct working cultures that were nurtured and protected in secure environments.[16] Radar technology brought together the perspectives and concepts of electronic engineering, fostered in a pre-war environment of radio and television development, with optical science. These blended cultures developed the hologram as the hybrid of seemingly incompatible ideas. Was advanced technology to be sequential and digital, or holistic and continuous, and which of these best fitted the spirit and goals of the age?

The orthodox analogue concept of the image was embodied in the established visual culture of photography. Over the previous century, photographers had developed a mature technology underpinned by scientific understandings. For them, the image was the result of a series of distinct operations. First, a lens focused an image of an illuminated scene on a photographic emulsion. The light altered the emulsion to create an invisible *latent* image, and a visible image could be recovered by chemical development, although its gradations of tone were reversed. In a darkroom, this negative image would then be projected from an enlarger to expose photographic paper, which then underwent its own chemical development to yield the photographic print. Importantly, these processes acted on the film and photographic paper *locally*—point by point, the photosensitive medium underwent the same sequence of exposure and processing. In short, a photograph was made by a series of steps carried out in parallel for every part of the image.

As outlined in Chapter 2, camera design and darkroom chemistry had become mature technologies. The amount of light in the scene and from the enlarger, the composition of the silver grains in the film emulsion and photographic paper, and the type and concentration of processing chemicals all altered the final image in complex but predictable ways.

[16] On the most highly-funded of these initiatives, nuclear science and technology, see Johnston, S. F., 'Security and the shaping of identity for nuclear specialists', *History and Technology* 27 (2011): 123–53, Johnston, S. F., *The Neutron's Children: Nuclear Engineers and the Shaping of Identity* (Oxford: Oxford University Press, 2012).

While the many variables important to each stage were well understood, the creation of a faithful image remained an art that was dispersed between different types of specialist: camera designer, photographic emulsion chemist, photographer and darkroom worker. Their skills overlapped only partially, and their shared expertise was needed to yoke this hybrid technology.

On the other hand, electronic imaging adopted a radically different approach. Television, developed from the 1920s, had forced engineers to rethink imaging from first principles. The first and most important shift was to think of images sequentially instead of holistically.

In a telephone line or radio channel, the tone and volume of sound are related to the frequency and amplitude of the radio wave. Images can be transmitted via such channels, but must first be broken down into discontinuous segments. To send an image, the first facsimile ('fax') machines, developed during the late nineteenth century, scanned the graphic into a series of stripes. Each stripe, showing brightness variations along successive lines through the image, became a time-varying electrical signal transmitted over a telephone line. When combined with radio transmission and half-tone printing, this technology became popular with telecommunications and news organizations during the early 1920s for 'wire photos'. In each implementation, the smoothly varying (analogue) voltage was punctuated periodically by the end of each stripe. Stringing together this periodic series of segments at the receiving end, the stripes could be recorded sequentially one by one to reproduce the scanned image.

Television developers extended this sequential concept further. A television signal subdivided moving pictures first as a horizontal stripe varying in brightness, then stacked this series of stripes vertically to create an image equivalent to one photograph or 'frame' of movie film. When one frame had been transmitted, the process repeated to construct the next image. A single electrical waveform, suitably divided into time-varying segments, produced the series of stripes and, from them, entire frames of successive television pictures to generate a two-dimensional moving picture. This reimagined *sequential* form of imaging paid dividends. A single telephone line or radio channel could transmit the time-varying video signal, making the process practical and cost-effective.

Electrical engineers were leaving behind the hybrid world of photography, but were developing a new hybrid of their own. Television represented a halfway house between the analogue and digital depictions of the world. The continually varying waveforms were analogue, but the way they were divided up into 'scan lines' and 'frames' added a discontinuous quality to imaging. The early mechanical-scanning scheme devised by John Logie Baird during the late 1920s divided the image into thirty stripes or lines stacked horizontally. By the mid-twentieth century, electronic scanning schemes employed several hundred vertically-stacked scan lines to build up the image, and modern high definition (HD) systems use over a thousand lines to generate a higher-definition image.

Apart from the television camera lens, the technology could be made entirely electronic, which opened possibilities for further manipulation the image. This all-electronic medium also meant that one breed of specialist, and one engineering culture, could dominate its development: from the 1920s, electrical engineering grew quickly. Television, then, encouraged a streamlining of concepts about imaging and a new hybridization of imaging specialists.

Such crossbred creations encouraged further subdivision of signals. The invention of the digital computer reinforced the utility of representing continuous measurements as a series of discrete values. Each value could then become an element of an arithmetic calculation and recombined with others in the string to yield an output that appeared continuous. This numerical approach had been used sporadically since late eighteenth century, especially for the calculation of orbits in astronomy and for generating mathematical tables for trigonometry and logarithms. It became more systematic and widespread, though, from the 1930s, when large teams of 'computers'—usually women—were marshalled. This human technology was essential for computing numerical tables of artillery trajectories and later in calculations important for designing the atomic bomb. It is not a coincidence that computing, governments and military applications became closely associated.[17] But, despite this convergence towards electronics and digital calculations, the notion of representing an image by a string of numbers had not yet been conceived.[18]

At the same time that electrical engineers were tentatively exploring digital calculations, they were making their work systematic in another way. Just as optics had been rendered mathematical, the electronic designs of the early part of the century were increasingly understood in theoretical terms. Unlike the first experimental fax transmissions, television picture quality was increasingly understood as a series of mathematical operations (a so-called *linear systems* approach), a powerful understanding that greatly simplified and expanded design possibilities. Practical improvements in radio and television were progressively supported by the mathematical field of 'communication theory'. By the early 1950s, then, there was a growing momentum for a new *systematic, serial, digital* approach favoured by engineers facing complex real-world problems.

On the other hand, most physicists disdained this rising culture and were based solidly in the analogue camp. They preferred to make approximations, where necessary, to obtain analytical solutions—equations that provided physical insights and that represented observation as a continuous process. Measurements of all kinds—voltage, brightness, volume, temperature—were part of a continuum. They were not perceived

[17] The breakdown of mathematical problems into step-by-step calculation algorithms was the key for these teams of human computers and for the digital computers that succeeded them. See Agar, J., *The Government Machine: A Revolutionary History of the Computer* (Cambridge, MA: MIT Press, 2003), Grier, D. A., *When Computers Were Human* (Princeton, NJ: Princeton University Press, 2005).

[18] The digital image so familiar today from computers, cameras and image scanners became available to consumers only from the late 1990s. Today image manipulation software routinely sharpens, brightens and colour-corrects the images from consumer digital cameras. Cameras themselves are thoroughly hybridized, incorporating automated versions of such sophisticated digital image processing to compensate lens aberrations, sensor defects and user motion.

as discrete values, and when they did appear in a numerical table it was understood that these were mere snapshots in time, artificial approximations that captured a pale shadow of reality.[19]

Gabor's extension of visual science forced together three uncomfortable communities: electrical engineers versed in the new stepwise linear analysis, physicists accustomed to sophisticated mathematical derivations with few immediate applications, and workaday microscopists trained in traditional optical methods. Where the engineers were trained to decompose images into stripes, picture elements or processing steps, the physicists sought to transform them mathematically, and microscopists focused on pragmatic techniques. These competing understandings marginalized Gabor's work, and had much in common with the earlier professional experiences of Gabriel Lippmann (Section 2.4), Ernst Abbe (Section 2.2) and Frits Zernike.

Abbe's powerful conception of how microscopes formed images was a reimagining of optics itself. His unintuitive mathematical concepts argued that a small aperture microscope or telescope could never produce images as sharp as a large aperture version because it 'filtered out' fine brightness fluctuations or *spatial frequencies* of the image. Microscopes did not merely magnify images; they transformed them in subtle ways. These ideas had close but as-yet unrecognized affinities with concepts in communication theory being developed by electrical engineers in the 1950s.

And the most recent example of the insularity of optical knowledge was the concept of the phase-contrast microscope, which won the Nobel Prize for Frits Zernike (1888–1966) in 1953. Like Abbe and Lippmann before him, Zernike's clever invention, making transparent objects visible under a microscope, also relied on mathematics that was belatedly recognized as a conceptual leap in the field of optics, yet readily grasped only by the next generation of engineers and physicists.[20] Attempting to combine incompatible disciplines, Gabor, like Lippmann, Abbe and Zernike before him, worked on the fringes of professional respectability.

These balkanized technical cultures were recreated in miniature at the Willow Run Laboratories (WRL) set up in Michigan after the Second World War. Sparsely scattered over a few hundred acres of land at the edge of an airport between Detroit and the university town of Ann Arbor, the squat white buildings and hangars of the Labs dated from before and during the War, when the Ford Motor Company had set up a car production

[19] The representation of mathematical expressions in tables had become popular as a calculating aid in the seventeenth century, particularly after the development of tables of logarithms. Quantitative measurement in the physical sciences, increasingly standardized from this time, relied on discrete measurements from which more general functional relationships could be inferred. Engineering practice amalgamated these approaches during the nineteenth century. See, for example, Johnston, S. F., *A History of Light and Colour Measurement: Science in the Shadows* (Bristol: Institute of Physics Publishing, 2001).

[20] Zernike observed, 'on looking back to this event, I am impressed by the great limitations of the human mind. How quick are we to learn, that is, to imitate what others have done or thought before. And how slow to understand, that is, to see the deeper connections. Slowest of all, however, are we in inventing new connections or even in applying old ideas in a new field' (Zernike, F., 'How I discovered phase contrast', *Nobel Lecture, Nobel Prize for Physics* (1953)).

plant and then a bomber plant. The unpromising environment housed a mixed population and hidden technical culture.[21]

Unlike pre-war university labs, the WRL research was goal-driven: the facility had been created to increase the income of its parent institution, the University of Michigan, by carrying out research and development contracts under generous funding from the American Department of Defense. The first contract, for an antiballistic missile system, was soon joined by others for battlefield surveillance, and incorporated ideas that had begun during the war. Infrared detection, microwaves and digital computers were studied at Willow Run. One particular hut on the edge of the airport flight line housed a small group developing what came to be called *synthetic aperture radar*, or SAR. Like facsimile imaging and television before it, SAR straddled the separate understandings and distinct cultures of electronics and optics. And from it, a decade later, came a revitalized and altered identity for the hologram.

4.4 FROM RADAR TO HOLOGRAMS

The hologram had a fresh start in this new environment, recognized by its small circle of developers as an efficient and powerful challenger to the first digital computers.

Electronic computing had developed rapidly during the Second World War to support urgent calculations. Two types had become popular with engineers: *analogue* computers, which modelled a mathematical relationship through an electronic circuit and smoothly varying voltages, and *digital* computers, which performed arithmetic via electronic pulses. Analogue computers were the first to yield practical results. With clever arrangements of electronic components and mechanical linkages, they could plot complex graphs, filter signals and reveal subtle mathematical relationships.

By the end of the war, however, digital computers were showing aptitude for more precise calculations, making them preferable for tabulation, finance and extended scientific calculations (Figure 4.1). Both analogue and digital computing promoted a burgeoning branch of mathematics. The new engineering fields were being enriched by a scientific orientation.

While Dennis Gabor reflected ruefully that holograms had gone into 'a long hibernation' in the late 1950s, Emmett Leith later argued that they 'went underground'.[22] Leith (1927–2005) had joined WRL in 1952 after a brief stint in the wartime Navy and gaining degrees in physics from a local university. Over the next decade, he and his colleagues

[21] As several of its members noted, much of their production was filed in classified reports and never published: 'many people's entire career was like that, doing very good work and getting no recognition, only a pay check' (Leith, E. N., D. Gillespie and B. Athey to SFJ, interview, 11 Sep 2003, Ann Arbor, SFJ collection; Emmett Leith quoted).

[22] Gabor, D., 'Holography, 1948–1971', in: Nobel Prize Committee (ed.), *Les Prix Nobel En 1971* (Stockholm: Imprimerie Royale P. A. Norstedt & Sonner, 1971), pp 169–201; quotation p. 176; Leith, E. N., 'The legacy of Dennis Gabor', *Optical Engineering* 19 (1980): 633–5; quotation p. 633.

Fig. 4.1 Postwar computing. 'A research engineer is seen setting up one of ENIAC's thirty-six units'. Electronic Numerical Integrator And Computer, University of Pennsylvania (Lilley, S., 'ENIAC, ASCC and ACE: Machines that solve complex mathematical problems', *Discovery* 8 (1), Jan 1947: 23–7, illustration p. 26, uncredited photographer).

gradually conceived a new hybrid of electronics and optics and, through it, gestated the first successful holograms.

Leith's task as a Research Assistant at WRL was to work on a new form of imaging radar. In wartime designs, radar pulses reflected off an object could determine its distance by measuring the delay time of the echo. The earliest schemes used an oscilloscope to display the reflected pulses as a graph. If the radar beam were swept through a circle, however, the distance and direction of objects from the transmitter could be mapped. Such swept radar reflections, familiar to cinema and television audiences since the War, depicted aircraft or ships as gradually fading spots, periodically updated, on a phosphorescent cathode ray screen.

The new form of radar was intended to provide much higher quality pictures from airborne radars—and not merely point-like objects but entire terrains. The system was intended to measure not just the echo, but also the frequency of the echo. The echo indicated how far away from the aircraft the reflecting object was, and its shifting frequency indicated a second dimension: how far ahead or behind the aircraft the object was via the Doppler effect. The idea was to record the radar echoes continuously on a magnetic tape, and then to analyse the recorded signal to decode the side-ways and forward-backward positions.[23]

Real-time analysis was inconceivable. The decomposition of the electronic signal into frequency components and time delays was much more complex than the processing in any existing imaging system, including television and conventional radar. Initial ideas were to play back the tape through an electronic circuit that would pass one frequency at a time, to generate a set of echoes corresponding to a line in front of, or behind, the aircraft. The radar picture would then be created by stacking those separate 'scan lines' much as television pictures were built up. But given the complexity of the filtering process, the scan lines would have to be recorded on photographic film rather than temporarily on a cathode-ray tube. To make matters worse, the filtering process threatened to be too crude to yield many scan lines, and so the resolution would be unimpressive.

This awkward hybrid of high-frequency radar transmission, tape recording and analogue electrical filtering forced the project engineers to consider alternatives. Calculations to process the recorded waveforms were effectively impossible: contemporary digital computers were too expensive, too limited in storage capacity and too slow to do the job. Emmett Leith was assigned the task of considering an optical approach. As Abbe and others had shown, lenses had mathematical properties relevant to the problem, and might offer possibilities for transforming the recorded data directly into an image.

Along with colleagues, Leith's dogged exploration of such an 'optical processor' eventually yielded a practical device in 1958. As the aircraft flew, the radar signal was recorded onto photographic film; the resulting transparency was then placed in an optical instrument to transform the complex pattern on it into a remarkably detailed image of the terrain below. With this miraculous apparatus came a rich intermixing of communication theory with modern optics, and the gradual realization that the radar recordings and optical processor could be understood as equivalent to Gabor's holograms and viewing apparatus. The two concepts came literally from different directions: Gabor had envisaged a new kind of microscopy based on extremely short-wavelength electron beams; Leith and his peers had developed imaging based on extremely long-wavelength radar waves. As Gabor, Leith and their colleagues began to identify deep parallels between electronics theory and optics during the 1950s, they separately linked the hologram and radar concepts to the mathematical foundations of modern optics. This merging of two leading-edge subjects is unusual at any time; to have it happen first in a postwar electrical

[23] Jensen, H., L. C. Graham, L. J. Porcello and E. N. Leith, 'Side-looking airborne radar', *Scientific American*, 237 (10), Oct 1977: 84–95.

company, and soon after in a security-conscious and goal-oriented contract research centre is all the more notable.[24]

Even more remarkably, there was, in fact, a third sequestered site where similar concepts were developing at about the same time. At the Vavilov Optical Institute in Leningrad, USSR—a domain as isolated as Willow Run Labs and equally dedicated to developing technologies for military application—an optical engineer pursuing an advanced degree evolved a concept that he dubbed 'wave photography'. The research of Yuri Denisyuk (1927–2006) had distinct origins, rediscovering and transforming some of the concepts that Gabriel Lippmann had explored a half-century earlier. Like Gabor and Leith, Denisyuk used mercury vapour lamps as light sources to make his holograms, although his arrangement was unique and—as much later realized—uniquely powerful.[25]

The isolated investigators shared other perspectives. Like Gabor and Leith, Denisyuk was enthused by his creation, but had limited aspirations for its future. Wave photographs, he suggested, might find some utility as components in sophisticated optical systems. As mirrored by the experiences of Gabor, Zernike, Lippmann and Abbe before him, the peers of Yuri Denisyuk found his concept unconvincing. Like Gabor's demonstrations, his experimental results were unremarkable, limited largely by his monochromatic lamps, and his explanations were partial and unintuitive. Unlike Gabor, Denisyuk had few connections and little status; his ideas languished and he returned to other more mundane technical problems until the mid-1960s.[26]

With a successful demonstration of the new synthetic aperture radar system in 1959, WRL engineers continued to explore applications of their insights. Notable directions were signal processing and image processing—both of direct interest to their military sponsors—along with episodic explorations of holograms. From 1960, Leith and a new co-worker, Juris Upatnieks (b. 1936), began to re-examine and extend Dennis Gabor's work methodically. As time permitted, they developed a practical solution to Gabor's

[24] By the late 1950s a number of optical scientists and electronic engineers were exploring links between their working cultures. Gabor and his collaborators at Associated Electrical Industries (AEI), coming from electronics backgrounds, sporadically applied communications concepts to their hologram research. A similar example was German physicist Adolf Lohmann, who was seeking a solution to Gabor's hologram problem. He identified a formal link between holograms and single-sideband (SSB) communication—a technique coincidentally becoming popular with ham radio enthusiasts just then. Straddling two cultures, he found the language and mind-sets of optical technologists and electronic engineers antagonistic; his own understandings proved difficult to communicate to either audience. Lohmann, A. W., 'Optical single side band transmission applied to the Gabor microscope', *Optica Acta* 3 (1956): 97–9; Lohmann, A. W. to SFJ, fax, 13 May 2003, SFJ collection.

[25] Denisyuk, Y. N., 'On the reflection of optical properties of an object in the wave field of light scattered by it', *Doklady Akademii Nauk SSSR* 144 (1962): 1275–8; Denisyuk, Y. N., 'Holography at the State Optical Institute (GOI)', *Soviet Journal of Optical Technology* 56 (1989): 38–43 Translated from *Optiko Mekhanicheskaya Promyshlennost* 56 (1989): 40–4; Denisyuk, Y. N., 'My way in holography', *Leonardo* 25 (1992): 425–30, Denisyuk, Y. N. to SFJ, email, 3 May 2003, SFJ collection.

[26] See Johnston, S. F., *Holographic Visions: A History of New Science* (Oxford: Oxford University Press, 2006), Chapter 3. Denisyuk's form of holograms had conceptual links with Lippmann's turn-of-the-century interference colour photographs and, when recorded years later with lasers and improved emulsions, allowed spectacularly faithful reproductions in full colour and three dimensions.

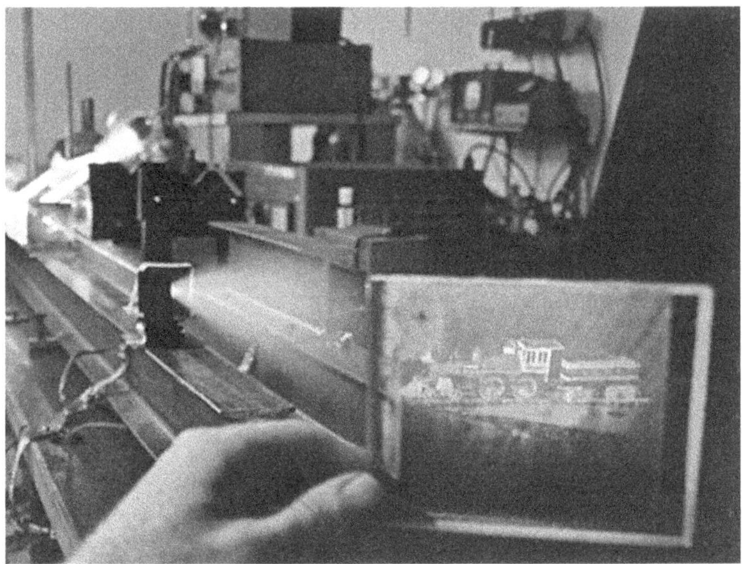

Fig. 4.2 Toy train hologram lit by an argon laser (hologram, 4 × 5 inches, silver halide on glass, recorded by Juris Upatnieks and Emmett Leith, Dec 1963; E. Leith photo 1965; A. Funkhouser collection and scan. The gaudily-painted toy train itself is now in the collection of the National Museum of American History (collection no. EM 330400; http://collections.si.edu/search/results.htm?q=EM+330400)).

experimental problems, finding a way of creating high-quality holograms for the first time. Their first successes, like Gabor's, used the best available pure-colour lamps: glowing mercury and sodium vapour, filtered by coloured glass filters. Because the lamps were not adequately monochromatic, their light sources limited them to two-dimensional images.

Lasers, invented as pulsed light sources in 1960, offered new possibilities: they produced 'coherent' light having a precise frequency and phase that was ideal for recording the interference fringes of holograms. But, in 1963, when Leith and Upatnieks were able to borrow a newly available *continuous* laser (which produced a constant intensity for creating and viewing the hologram), they were finally able to make *three*-dimensional images with similarly high quality (Figure 4.2).[27] By that Christmas, the two investigators were showing the first successful holograms of three-dimensional scenes, lit by the eerie light of a red laser beam, to their colleagues at Willow Run.[28]

[27] The first ruby laser, demonstrated by Ted Maiman of Hughes Research Laboratories, generated pulses of deep-red light. By 1966, this had been further developed to make successful 'pulsed' holograms. The red-light helium–neon laser, developed at Bell Telephone Laboratories in 1962, was the first type used to make holograms, followed commercially from 1965 by the more intense blue-green light of the argon laser.

[28] Leith, E. N. and J. Upatnieks, 'New techniques in wavefront reconstruction', *Journal of the Optical Society of America* 51 (1961): 1469 and Leith, E. N. and J. Upatnieks, 'Reconstructed wavefronts and communication theory', *Journal of the Optical Society of America* 52 (1962): 1123–30; Leith, E. N. and J. Upatnieks, 'Wavefront reconstruction with diffused illumination and three-dimensional objects', *Journal of the Optical Society of America* 54 (1964): 1295–302.

5

Holograms as Magic

> *They say 'you won't believe it after you've seen it'. Holography (pronounced 'hall-ography') is a new extension of photographic methods made possible by the use of lasers. The name holography is Greek for 'whole picture', and the process produces three-dimensional images (without the use of viewing spectacles or other devices) which are an exact replica in space of the wavefronts of light emanating from the original scene. One can move his head from side to side (or up and down) and actually see behind objects in the foreground. If you are interested in the vanguard of science this will be* <u>*the*</u> *program for you. As always, students and visitors welcome.*[1]
>
> Fort Worth, 1966

The campus of the University of Michigan, in the small town Ann Arbor, seems a million miles from the Willow Run Labs (WRL). The attractive modern buildings, trimmed lawns and ample boulevards of North Campus appear to share nothing with the white huts and utilitarian services on the windswept airport land just fifteen miles away. Yet this is a superficial assessment: many of the same engineers and scientists, in fact, worked in both locales. And most importantly for this story, it was in these places that the hologram made its public debut.[2]

The Willow Run engineers who cast the hologram as the heart of data processing—a kind of optical CPU—gradually conceived another role for it. This new identity was so compelling that university administrators decided to move the team to a more accessible location, and to make their expertise, and their holograms, available to business interests and students. Leith and Upatnieks moved to the new Institute of Science and Technology (IST) being built at the University of Michigan in 1964.

The segue involved more of Ann Arbor. Willow Run created a diaspora of technical workers. Some of them founded new companies in the town. One of those companies, the grandly named Conductron Corporation, had set up shop in 1960 to work on Department of Defense contracts and to commercialize spin-offs. At Conductron and other offshoot firms, the hologram was steered toward imagined audiences and planned application niches.

[1] 'Holography – photography in three dimensions at Indiana Institute of Technology', *The Announcer*, 7 (2), Oct 1966: 1.

[2] Leith, E. N., 'Electro-optics and how it grew in Ann Arbor', *Optical Spectra*, May 1971: 25–6.

Through the mid-1960s, holograms were seen mainly by engineers and reporters at conferences or private demonstrations. As a result, these interpreters were as important in framing the nature of the new invention as the creators themselves. And the easiest fit for their interpretation was the ancient appeal of magic.[3]

5.1 MODERN ILLUSIONS

The three-dimensional hologram and the rapidly developing lasers to make it were unanticipated spin-offs of classified research. Defence contractors, government funders and a growing body of engineers became enamoured with the potential of holograms.

Much of this potential was secret. The primary application of the Willow Run techniques was for image processing: transforming, revealing and clarifying data. For Leith's Willow Run colleagues, the new field of optical processing offered challenging problems and powerful capabilities, but they were veiled by Department of Defense secrecy.

Emmett Leith and Juris Upatnieks had been working on ways of improving holograms for three years and publishing their progress in specialist journals at a steady pace, but Leith had been chagrined on three counts. First, he and his co-workers were muzzled. Colleagues working on the electronics portion of side-looking radar had been able to publish articles on it in open journals, but the optical contributions—the heart of the analysis system—could not be revealed to the public.[4] Second, the powers of optical processing for military applications were being overshadowed by early forms of digital processing of images in the nascent space programme. The techniques provided a way of sharpening, filtering and even automatically selecting targets in images from cameras, radars and infrared sensors. Only rarely did such applications escape the lab to intrigue the wider public. Computer-based methods to achieve this were still twenty years away, but were already more discussed in popular media.[5] And third, a recently-joined senior colleague in physics, Professor George Stroke (1924–2007), was stealing their thunder

[3] On the survival of magic from the early Enlightenment into modern contexts, see Davies, O., *Witchcraft, Magic and Culture, 1736–1951* (Manchester: Manchester University Press, 1999).

[4] See, for example, Pearson, J., 'The Army's new eyes: U-M Scientists develop far-sighted radar unit', *Detroit Free Press*, 20 Apr 1960, 2; Dodd, P., 'A spy in the sky! Army unveils radar camera: it shoots enemy thru clouds of smoke', *Chicago Daily Tribune*, 20 Apr 1960, 3.

[5] Little of the optical image processing work was reported publicly, and there was competition through the 1960s with investigators exploring early methods of image enhancement by computer. The Jet Propulsion Laboratory (JPL), tasked with developing digital imaging techniques for the space programme, received wide acclaim for enhancements of the first lunar orbiter spacecraft images, which had been obscured by scan lines and low contrast. Within a decade, the analogue optical methods had lost ground to digital processing techniques in medical and planetary imaging, but were still employed for their original high-throughput application of radar image processing. For examples of the battle for publicity, see Jet Propulsion Laboratory Office of Public Information, 'JPL computer process brightens Surveyor moon pictures', press release, 9 Aug 1966, http://www.jpl.nasa.gov/news/releases/60s/release_1966_0402.html, retrieved 16 Jul 2014; 'Holography: clearing the image', *Time*, 91 (12), 22 Mar 1968: 69; de Plante, R., 'How computers clear up fuzzy pictures', *Deseret News (Arizona)*, 24 Sep 1969, A11; Stroke, G. W., 'Sharpening images by holography', *New Scientist* 51 (1971): 671–4.

with claims about the invention of holograms.[6] Hologram research offered an avenue for recognition and a potential field of research 'big enough to spend a lifetime in', as Leith later reflected.[7] Consequently, both Leith and Stroke promoted their work not just within their cultures of engineering and physics, respectively, but in newspapers and magazines, too.

Leith's and Upatnieks' first conference presentations and journal papers, describing their experimental solution to Dennis Gabor's flawed techniques and a theoretical explanation in terms of communication theory, passed with little notice. But in late 1963, they had their first taste of celebrity, at least among their peers. The American Institute of Physics (AIP), newly conscious of the publicity value of scientific research, picked up on an angle that Dennis Gabor had tried to exploit in 1948. Gabor had impressed his postwar audiences with the ability to scramble a photographic image, and then to 'reconstruct' a recognizable image from the confused hologram. By the summer of 1963, Leith and Upatnieks could improve this conjuring trick dramatically: *their* holograms could reveal a sharp image with excellent rendition of tones from an emulsion that appeared to have recorded featureless grey mush. It still required apparatus—a glass hologram mounted on an optical bench and lit by a monochromatic lamp or laser—but the awe was infectious. As Leith recalled:

> It was most fascinating, because here you had a piece of film that had nothing but garbage on it, or very fuzzy images at best, and then you looked downstream where they came to a focus, and there you saw a real, nice, sharp image, and there was nothing producing it—there was an image but no optical elements—kind of like a grin without a cat by Lewis Carroll's analogy, you might say.[8]

Even when that enchantment was drily reported in their conference paper, the AIP administrators found the ideas so compelling that they issued a press release and a catchy label. They replaced Gabor's cumbersome 'wavefront reconstruction' with a more evocative label, 'lensless photography'. Over the next two years, that term packaged the idea and framed public understandings.[9]

Just as Gabor had briefly captured the attention of the *New York Times,* Leith and Upatnieks attracted—but this time captured—media attention. In early December 1963, both the *Times* and the *Wall Street Journal* interviewed Leith by telephone. Both papers published careful stories, in some cases alongside the developing coverage of John Kennedy's

[6] For four years Leith and Upatnieks were challenged by University of Michigan Professor George Stroke (1924–2007), who claimed priority in conceiving holography for diffusely-reflecting three-dimensional scenes. His misleading accounts of the field secured the Nobel Prize for holography to Dennis Gabor alone (Johnston, S. F., 'Telling tales: George Stroke and the historiography of holography', *History and Technology* 20 (2004): 29–51).

[7] Hecht, J., 'Applications pioneer interview: Emmett Leith', *Lasers & Applications*, 5 Apr 1986: 56–8.

[8] Leith, E. to S. F. Johnston, interview, 22 Jan 2003, Santa Clara, CA, SFJ collection.

[9] American Institute of Physics, 'Press release: Lensless optics system makes clear photographs', 5 Dec 1963, p. 2. See also Johnston, S. F., 'Absorbing new subjects: holography as an analog of photography', *Physics in Perspective* 8 (2006): 164–88.

recent assassination. The stories, picked up by other newspapers, stressed the novelty that the AIP press release had isolated. This new imaging process was a kind of scientific sorcery: a 'camera without a lens' which could clarify a smudged or blurry negative, or perhaps even magnify hidden details. The claims magnified those made by Gabor fifteen years earlier and, this time, media interest was sustained.[10]

Despite their background in a high-security field, the investigators proved adept at enticing audiences. Their skills were no doubt honed at WRL, where selling the promise of new technological possibilities to government funders had risen to a high art. Research and Development contracts, not peer recognition, were the bread-and-butter of such centres. It is notable that the first wave of hologram producers and promoters came almost exclusively from the technicians, scientists and managers at WRL.[11]

Timing was also a factor. Just as this first media interest was peaking, Leith and Upatnieks were obtaining their first successes with *three*-dimensional holograms, the result of a year's dogged effort to tame the newly-available lasers. According to Upatnieks' recollections, their attempts to demonstrate three-dimensionality had been triggered by the initial press attention itself.[12]

Upatnieks and Leith had used small photographic plates just large enough for a laser beam or single eye to look through, but now tried wider plates; Leith recounted:

> we cut them to this length so two eyes could get in, and sure enough, they were three-dimensional as prescribed. But the realization was unbelievable; just seeing it there, seeing it happen, was most astonishing.[13]

And Upatnieks recalls, 'once we succeeded, we were fascinated by the reality of the image and spent hours looking at it, and showed it to our colleagues' (Box 5.1).[14]

As a handful of colleagues and reporters began to drift into their labs through the winter of 1963, Leith and Upatnieks could share the excitement about the new form of magic. Engineering visitors from the Ann Arbor region—WRL co-workers, University of Michigan scientists and occasionally friends of employees—were the first witnesses. Most marvelled at the laser-lit hologram of a toy train engine, the hobby product of a WRL technician, borrowed because its weight encouraged it to stay still for the ten-minute exposures.

[10] See 'Lensless photography uses laser beams to enlarge negatives, microscope slides', *Wall Street Journal*, 5 Dec 1963, 28. Leith and Upatnieks themselves adopted the term in their first conference presentations, e.g. Upatnieks, J. and E. N. Leith, 'Lensless photography', presented at *SPIE Photo-Optics Workshop Meeting*, Los Angeles, 1964. An additional reason for their adoption of the label was that the term 'holography' had been appropriated by their rival George Stroke. Owing in large part to Stroke's networking in the physics community, 'holography' dominated as the name of the new field after late 1965.

[11] See Johnston, S. F., *Holographic Visions: A History of New Science* (Oxford: Oxford University Press, 2006), Section 6.3.

[12] Upatnieks resumed experiments in early December that had not succeeded that spring, 'since reporters quizzed us about making holograms of reflecting objects and their 3-D image qualities. We had not done that yet, but these properties were obvious to us' (Upatnieks, J. to SFJ, email, 9 Jan 2005, SFJ collection).

[13] Leith, E. N. to SFJ, interview, 22 Jan 2003, Santa Clara, CA, SFJ collection.

[14] Upatnieks, J. to SFJ, email, 24 June 2003, SFJ collection.

Box 5.1 Viewing Leith–Upatnieks holograms

When lit by a diverging laser beam, a Leith–Upatnieks (L-U) or 'laser transmission' hologram recreates ('reconstructs') an image that appears behind the hologram plate. The experience is exactly like looking at the original scene through a window. The image is life-sized and may be as much as several metres deep.

With suitable recording arrangements, an L-U hologram may also reveal an image that passes *through* the plate or hangs in front of it. This variant is known as an 'image-plane transmission hologram'.

The laser light adds a subtly grainy or mottled appearance to the scene ('laser speckle'), and shifts quickly as the viewer changes position.

The orientation and spread of the laser beam must match the arrangement that was used when originally recording the hologram to avoid distorting this so-called 'virtual image' of the scene.

When lit by an *unexpanded* laser beam, the hologram can focus an image onto a screen. The viewing experience is much like observing the original scene on the screen of a traditional camera. Again, the orientation of the laser beam and hologram plate are critical. The appearance and viewpoint of this 'real image' of the scene depends on which portion of the hologram is intercepted.

If the laser is replaced by a white light source, the images in both cases will be smeared into an unrecognizable blur.

Given the security-conscious and relatively isolated airport labs, casual visitors were uncommon. For three months, the private viewings dampened publicity. The second media exposure, another conference paper to optics specialists, provided the spark for public interest.

In early April 1964, Upatnieks presented a paper at an optics conference session devoted to their main research strength (information processing by optics) but aimed at the very different application of three-dimensional imaging.[15] As Leith remembered it, demonstration trumped description. Showing the hologram of the model train mystified the viewers who queued to see it:

> Even though most of them were optical scientists, they did not understand what was going on. They assumed it was a projection system done with mirrors, and that the little toy train was hidden somewhere.[16]

[15] Upatnieks, J. and E. N. Leith, 'Lensless three-dimensional photography by wavefront reconstruction', presented at *Optical Society of America Spring Conference*, Washington DC, 1964; the expanded version was published as Upatnieks, J. and E. N. Leith, 'Lensless, three-dimensional photography by wavefront reconstruction', *Journal of the Optical Society of America* 54 (1964): 579–80.

[16] Hecht, J., 'Applications pioneer interview: Emmett Leith', *Lasers & Applications*, 5 Apr 1986: 56–8; quotation p. 56.

When Leith and Upatnieks published an expanded paper in the principal American optics journal in December 1964, the AIP again provided a press release. The result was sustained international visibility. If the first AIP press release had launched a signal rocket to capture brief attention for lensless photography, the second opportunity for publicity kicked three-dimensional holograms into a stable orbit.[17]

5.2 PORTRAYING PERPLEXITY

The strange properties of holograms were difficult to get across. Through 1964 and 1965, presentations by Leith and Upatnieks to engineers and scientists communicated the unintuitive characteristics and fertile possibilities of this bizarre imaging technology. At each venue, their demonstration holograms generated surprise and curiosity among their peers, and even disbelief among engineers from more distant fields.

At a time when few professionals had encountered lasers, seeing the laser-lit hologram and the unexpectedly empty space behind it suggested technological magic as much for the experts as for wider publics. Even the *essence* of the technology was difficult to assimilate: was this a variant of photography, modern communications like television, or a unique property of laser lighting?

The first newspaper article, the result of a demonstration and interview by Emmett Leith, explained:

> The light bounces from the subject, into a mirror and onto film. The result is a transparency that looks to the eye like a buttermilk sky. But when laser light is played upon it, the original scene takes shape in three dimensions.[18]

The problem was to communicate the magical visual effect with only words. In those early years of holograms, remarkably few photographs attempted to capture their appeal. Part of the reticence was technical. Holograms had to be lit by laser light, and photographs tended to emphasize speckle (a side-effect of interfering light waves that appeared as an unusual shimmer to the viewer's eye, but as ugly grain when photographed). Snapshots were awkward in the dark labs. And the striking reality of holograms also restrained photographers: at best, they could capture a scene that looked genuine, but which, of course, sacrificed three-dimensionality. Leith, Upatnieks and their peers typically reproduced slides or journal illustrations that showed a black-and-white image of the reconstructed object.

For wider audiences, science was recast in more appealing dimensions. Newspapers and magazines, when they included illustrations at all, often focused on a dimly-lit scientist crouching over his apparatus. The mystery of modern science in its secure

[17] American Institute of Physics, 'Press release: Objects behind others now visible in 3-D pictures made by new method', 25 Oct 1964.

[18] Lutz, W. W., 'New discoveries at Michigan universities', *The Detroit News—The Passing Show*, Sunday, 23 Feb 1964, 1.

labs was often enough to provoke curiosity. Cold War intrigues, espionage and covert technologies were recurring themes in contemporary news reports, novels, cinema and television.[19]

The laser, with its pencil-thin beam, channelled further intrigue. James Bond in *Goldfinger* (released autumn 1964) had been threatened by fictional high-power laser beam, again linking secret research, scientific advance and adventure. Indeed, the first newspaper report on the Willow Run work mentioning holograms—a mere two column inches—was subtitled 'Death Ray in Laboratory'.[20] Reporters got across the sense of novelty and speed of discovery: the notion of the *breakthrough* pervaded popular understandings of science. The term had subliminal links with military advance, having been popularized during the trench warfare of the First World War.

Photographs, rather than diagrams, introduced the new invention. Views of optical set-ups proved popular, often with the laser beam traced out through cigarette smoke. If exposure was well-controlled, the photographer might capture a distinguishable reconstruction through the small illuminated hologram window. Perhaps reporters sensed that labelled illustrations would communicate the wrong message: an arcane and incomprehensibly complicated new science, rather than a wonder-filled and mysterious technology. Verbal and textual descriptions protected that magic and saved it for an eventual lived experience. In any case, diagrams illustrating the physical principles began to appear in the popular press only some eighteen months later.[21]

For reporters, confused themselves, reporting the experiences of other witnesses was a reliable fall-back. Focusing on the bemused reactions of the experts allowed them to communicate the degree of unfamiliarity that holograms created:

> Most people find their first look at a hologram an astonishing encounter. 'There's a characteristic delighted giggle that seems to go with the first experience', says Thomas Vogl, director of Westinghouse's hologram research program. Even scientist and engineers, who usually know that holograms exist, are fascinated. 'Last year at the convention of the Institute of Electrical and electronic Engineers in Washington', says an official of Conductron Corp., a company involved in holographic work, 'we had a hologram on display. Everybody spent his time looking at the hologram and ignoring the rest of the booth'.[22]

[19] For example, *Danger Man* (ITV, 1960–62, 1964–68; retitled *Secret Agent* on CBS); Soviet downing of American U2 spy aircraft (May 1960); Bay of Pigs invasion (Apr 1961); Berlin wall construction (Aug 1961); *The Avengers* (1961–69); Cuban missile crisis (Oct 1962); *The Manchurian Candidate* (Dir. John Frankenheimer, 1962); *From Russia With Love* (novel by Ian Fleming, 1957; film dir. Terence Young,1964); *The Man From U.N.C.L.E.* (NBC, 1964–68); *Dr Strangelove* (Dir. Stanley Kubrick, 1964); *The Spy Who Came In From The Cold* (Dir. Martin Ritt, 1965); *The Ipcress File* (Dir. Sidney J. Furie, 1965); *Get Smart* (NBC, 1965–69; CBS, 1969–70); *I Spy* (NBC, 1965–68); *The Quiller Memorandum* (Dir. Michael Anderson, 1966); *Mission Impossible* (1966–73); *Topaz* (novel by Leon Uris, 1967; film dir. Alfred Hitchcock, 1969).

[20] *Goldfinger*, Dir. Hamilton, G., United Artists (1964); Lutz, W. W., 'New discoveries at Michigan universities', *The Detroit News—The Passing Show*, Sunday, 23 Feb 1964, 1. See also Slayton, R., 'From death rays to light sabers: making laser weapons surgically precise', *Technology and Culture* 52 (2011): 45–74.

[21] Leith, E. N. and J. Upatnieks, 'Photography by laser', *Scientific American*, 212 (6), Jun 1965: 24–36.

[22] Gilmore, C. P., 'Those incredible holograms: Amazing "frozen light waves" give first true 3D pictures', *Popular Science* 189 (1966): 78–91, 174–5; quotation p. 82.

> It is the most startling thing I have seen', said a nuclear scientist from Argonne National laboratory. A navy captain said, 'It's fantastic. I don't see how it can be done'. Many of those who view the picture ask to touch the photographic plate to assure themselves that it is indeed flat and that the Michigan exhibitors are not trying to pull a fast one.[23]
>
> The General in Washington thought he saw a chessboard. He reached out to touch one of the pieces. But the board was not really there. He was seeing a three-dimensional projection of the chessboard produced by the astonishing new technique of "holography.[24]

Nearly all of those expert witnesses were technologists. Yet the disorientation, surprise and enchantment of holograms vied with rational explanations. Like any magic show, the viewing experience was one of childlike wonder:

> In a Boston hotel suite recently, a few dozen normally sedate scientists and engineers were playing with a toy locomotive, a toy train-conductor and other such items . . . The toys had been 'reconstructed' by a technique that looks simple, yet is one of the most sophisticated developments in modern science.[25]

Even the Boy Scouts promoted holograms. Along with a sober and accurate survey of the field in an article in *Boy's Life*—including technical limitations as well as potential—its author sought to recall the adventure of boys' novels of an earlier generation:

> The use of holograms by museums raises an interesting possibility for a mystery thriller. A few years ago the movie Topkapi kept viewers on edge with the story of some thieves who stole a valuable jewel from the former sultan's palace in Istanbul [. . .] They succeeded, but imagine their distress if they had gone to all that trouble only to discover that the jewel was actually a very realistic hologram.[26]

Communicating scientific mystery was a rare privilege. Paul Kirkpatrick, who had corresponded with Dennis Gabor and supervised Albert Baez in exploring holograms during the late forties and early fifties, abandoned explanations to give a sense of their sorcery:

> Light waves recorded on a hologram are like scrambled eggs . . . and it's almost a law of nature that you can't unscramble eggs. Yet that what we can do with a hologram. It's almost unbelievable.[27]

Such third-hand experiences of the spectacular and unbelievable intrigued popular audiences, but opportunities to view holograms remained rare.

[23] 'Photo shows work of new optic system: viewers call exhibit startling, fantastic', *Chicago Tribune*, 27 Oct 1965, 36.

[24] Davy, J., '3-D pictures in the round', *Observer*, 15 May 1966, 13.

[25] Novotny, G. V., 'The little train that wasn't', *Electronics*, 37 (30), 30 Nov 1964: 86–9; quotation p. 89.

[26] Moffat, S., 'Science: the miracle of the laser beam brings new dimensions to the world of photography', *Boy's Life*, Jun 1969: 19.

[27] Gilmore, C. P., 'Those incredible holograms: Amazing "frozen light waves" give first true 3D pictures', *Popular Science* 189 (1966): 78–91, 174–5; quotation p.80.

5.3 DISTILLING THE ESSENCE OF HOLOGRAMS

Understanding holograms required fitting them into a conceptual schema. More than we often appreciate, this framing is not natural and automatic: it is shaped by culture. But holograms were too unfamiliar and resisted categorization.

The two AIP press releases coached readers to conceive the hologram as a new kind of photographic technique having novel qualities. The first announcement focused not on the hologram itself, but on a 'camera-like device' or 'an optical system in which a sharp, clear image is formed entirely without the use of lenses'. As to the hologram, it was akin to a 'blurred photographic negative'. The punch-line of the publicity was that this new form of photography empowered by laser light did not require lenses (Figure 5.1).[28] The second press release a year later revealed the possibility of three-dimensional pictures with heightened reality. The inventions were black-boxed as a kind of magicians' trick.[29]

The task of reopening that box to summarize its magical qualities proved frustrating, though. Waves of science writers strained to categorize and explain holograms, and fell back on a series of inadequate metaphors.

Among the earliest descriptions of holograms were hints of upgraded video technology for remote viewing:

> The train wasn't really there at all. But if you stood in exactly the right place and looked into a piece of equipment you would have seen it, real as life. The toys had been 'reconstructed' by a technique that looks simple, yet is one of the most sophisticated developments in modern science.[30]
>
> At the NEC exhibit, long lines of scientists saw a chess board with five men, a toy railroad scene and a model of an army tank. The objects were really back at the university's laboratory in Ann Arbor.[31]

The spectacle was thereby tamed: a baffling science could be reimagined in terms of more familiar communication technologies, or even more conventionally as a stage trick with mirrors. Such strategies of description paralleled those that had introduced interwar television and postwar 3-D movies.[32]

[28] 'No-lens picture-taking devised by U M men', *Ann Arbor News*, 5 Dec 1963, 31; Deschin, J., 'No-lens pictures: photographic technique employs light alone', *New York Times*, 15 Dec 1963, 143; 'Lensless laser photography.', *Popular Electronics*, Mar 1964: 60; '3-D lasography – the month old giant', *Laser Focus*, 1 Jan 1965: 10–15.

[29] American Institute of Physics, 'Press release: Objects behind others now visible in 3-D pictures made by new method,' New York, 1964, p. 1.

[30] Novotny, G. V., 'The little train that wasn't', *Electronics*, 37 (30), 30 Nov 1964: 86–9; quotation p. 89.

[31] 'National Electronics Conference, Chicago', *Pontiac Press*, 28 Oct 1965, 33.

[32] On the other hand, the notion of 'remote viewing' achieved a greater cultural visibility when it became associated with parapsychology research in the late 1970s. Links to popular understandings of holograms developed from the 1990s, when an American military study, the Stargate Project, was declassified and increasingly mythologized.

Fig. 5.1 Lensless photography in the comics (*Ann Arbor News*, 7 June 1964; courtesy of A. Spilhaus estate. A. Funkhouser collection and scan). *Our New Age* was syndicated to 121 American and foreign newspapers between the late 1950s and early 1970s. Athelstan Spilhaus (1911–1998), a South African geophysicist, sometime Dean of the University of Minnesota's Institute of Technology, UNESCO ambassador and board member of Science Service, recalled starting the comic because he 'felt dejected about the state of the human spirit in the United States at that time in the late fifties. And I remember saying publicly in the newspapers, "America needs a first".' He described his promotion of spaceflight, colonization of oceans and other grandiose projects of the future—'military support, free enterprise bases'—as intending to encourage entrepreneurship and reduce international tensions (Spilhaus, A. 'Oral history interview by Ronald E. Doel' (College Park, MD: Niels Bohr Library and Archives, 1989)).

A more stretching metaphor conflated the ideas of 'freezing' light with what was later dubbed, in the era of video-recorders, 'time-shifting'. The original press release had explained, 'The process can be thought of as capturing and storing the light rays and releasing them at some later time, whereupon the imaging process is carried to completion'.[33]

Subsequent science writers echoed:

> Basic to holography is the idea of freezing light interference patterns in a moment of travel; then, at any convenient time, freeing them to continue their journey. As they are released from the hologram, the light waves recreate the scene just as it existed, although the objects have been removed.[34]

This concept struck familiar cultural chords: for sixty years, the public had been buying sound recordings to play back musical entertainments at home. For the past fifteen, tape recorders had personalized this power to capture and release sound, and within the current decade stereophonic technologies were beginning to recreate life-like soundscapes via long-play vinyl recordings and FM radio.

But some properties mapped more satisfyingly onto a *trompe l'oeil* or conjuring act than onto consumer technologies. The holograms of Leith and Upatnieks generated visual confusion and reproduced stage levitation routines, séances and vanishing tricks—seemingly hanging in space, being revealed as ethereal when sought with a hand, or disappearing instantly when the laser was turned off. The magicians of vaudeville and contemporary variety television had trained generations to appreciate these tricks as entertainments to be enjoyed but not analysed.[35]

Still other properties of holograms, though, defeated analogy.

Some were appreciated mainly by photographic experts. Juris Upatnieks bemused early engineering audiences by noting that a contact copy of a hologram generated not a negative image, but another identical three-dimensional reconstruction, and that the contrast of the fringes on the hologram plate had little bearing on the dynamic range and tonal gradations of the reconstructed image.[36]

But the property dubbed 'real 3-D' had greater impact. 1960s audiences familiar with, but dismissive of, Victorian stereo views and 3-D motion pictures could appreciate that these visual experiences were imperfect. They required viewing apparatus (stereoscopes or glasses) and could produce disorientation and eyestrain when the cameras were not carefully aligned. Stereography was also limited to a fixed viewpoint. It provided a strong

[33] American Institute of Physics, 'Press release: Objects behind others now visible in 3-D pictures made by new method', 25 Oct 1964.

[34] Reinert, J., 'The picture that hangs in midair', *Science Digest* 60 (1966): 30–4; quotation p. 30.

[35] Berry, J. R., 'Holography: 3D magic in mid-air', *Popular Mechanics* 129 (1968): 104–7, 206.

[36] Leith, E. N. and J. Upatnieks, 'Wavefront reconstruction with continuous-tone transparencies', *Journal of the Optical Society of America* 53 (1963): 522.

sense of depth, but did not reproduce *parallax*—the ability to look around objects to see others in the background.[37] As reporters enthused,

> When you looked at the hologram, illuminated from behind by a gas laser, you saw the train and conductor toys right there on the table, in three dimensions. If you wanted to see what was behind the little man, or in front of the toy locomotive, you simply moved your head to see them. No need for viewing glasses, double images or squinting.[38]
>
> The scene in the Michigan lab was real 3D—without special glasses. By lifting my head, I could look down on a [toy] tank's turret; when I stooped, I was level with the treads. I could move my head and look around things. And the closer I got to the window, the wider my field of view became.[39]

In practice, this ability to bob and weave, to examine a hologram scene as though peering through a window, was compelling. The irritation, eyestrain and restrictions of perspective were replaced by a faithful and undistorted collection of views.

A second novel property was that—exactly like a real scene—holographic images required the viewers' eyes to focus at different depths. This subtle requirement was not shared by previous stereoscopic technologies. When viewing a stereo view or 3-D movie, the eyes focus at a fixed distance onto the printed cardboard picture or the display screen, rather than on objects in the scene that appear nearer or farther away. This is one reason why such systems can often seem inadequate or fatiguing. The need to focus a camera at different portions of the holographic image demonstrated that the recording had unlimited 'depth of field'. So unusual was this property that Juris Upatnieks highlighted it via photographs in their first papers (Figure 5.2). The older technology of stereoscopes provided some depth cues, but not all. They encouraged eyes to point in the right direction but to focus unnaturally, and they simulated an immobile head. Holograms, but contrast, behaved like windows onto a realistic but intangible world. Yet, unless witnessed in person, such subtle distinctions proved difficult to communicate.[40]

A third and more baffling optical property was the encoding of such realism within a featureless glass plate—the 'buttermilk sky' described by Emmett Leith's first newspaper reporter, as 'a dirty windowpane – about as fascinating as a snowball to an Eskimo' by another, and 'a montage of globs on a plate' by yet another (Figure 5.3).[41] Such

[37] de Closets, F., 'Le laser «voit» l'invisible: Révolution en optique: le relief total en photographie', *Sciences et Avenir* (1964): 162–7, 212.

[38] Novotny, G. V., 'The little train that wasn't', *Electronics*, 37 (30), 30 Nov 1964: 86–9; quotation p. 86.

[39] Gilmore, C. P., 'Those incredible holograms: amazing "frozen light waves" give first true 3D pictures', *Popular Science* 189 (1966): 78–91, 174–5; quotation p. 79.

[40] For example, Smith, C. W., '3-D or not 3-D', *New Scientist* 102 (1984): 40–4. For more on the qualities of stereoscopic vision, see Appendix.

[41] Gilmore, C. P., 'Those incredible holograms: Amazing "frozen light waves" give first true 3D pictures', *Popular Science* 189 (1966): 78–91, 174–5; quotation p. 78; Edson, L., 'Box cameras were never like this', *Popular Mechanics* 126 (Jul 1966): 67–9, 171, 3; quotation p. 173; Gabor, D. to W. L. Bragg, letter, 7 July 1948, IC GABOR EL/1.

Fig. 5.2 Photographs focused at different depths of hologram scene (originally published in Leith, Emmett N. and Juris Upatnieks, 'Wavefront reconstruction with diffused illumination and three dimensional objects', *Journal of the Optical Society of America* 54 (1964): 1295–1302; scanned photos courtesy of J. Upatnieks).

experiences were magical. Holograms were liminal objects: featureless panes of glass which, when lit by a laser, became a window onto another world. These modern crystal balls remained clouded for those untrained in the magic.

And a fourth intriguing quality of holograms was closely related to the third. It fitted not a conjuring trick, but a philosophical riddle: a small part of a hologram was able to reproduce the whole scene.[42] The earliest accounts of holograms linked the featureless appearance to this miraculous characteristic:

> An unusual property of the transparency is that the whole or any part of it contains the entire picture. Tear it up and any fragment of it will reveal the total picture under laser light.[43]

> Because there are no lenses, each point on the object is recorded all over the photographic plate. So you can take a hologram and cut it in half—or in a dozen pieces—and each piece will still show the entire object, from a slightly different point of view, with only a little loss in sharpness.[44]

[42] This odd characteristic can nevertheless be made familiar by imagining the Leith–Upatnieks (L-U) hologram as a window rather than a photograph. We can view an outdoor scene through a whole window, but we are equally able to view the scene by sighting through a small part of the window—between venetian blinds or through a peep-hole, for example. Just like a window, an L-U hologram intercepts a complex wavefront of light, not an image. And, like a window, a masked hologram provides a view of the whole scene, but limited to the particular viewpoint of that window element. Our bafflement might be more acute if we imagine the hologram broken into pieces, with each piece able to render the entire scene, but should not be: every shard of the L-U hologram acts like a viewing hole in a window.

[43] Lutz, W. W., 'New discoveries at Michigan universities', *The Detroit News—The Passing Show*, Sunday, 23 Feb 1964, 1.

[44] Novotny, G. V., 'The little train that wasn't', *Electronics*, 37 (30), 30 Nov 1964: 86–9; quotation p. 87.

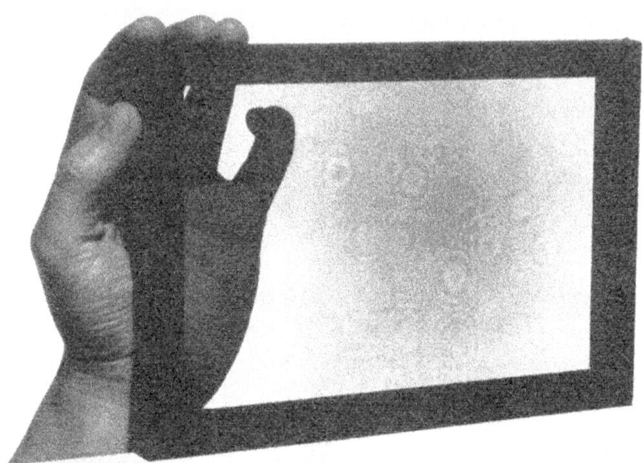

Fig. 5.3 Leith–Upatnieks, or laser transmission, hologram viewed without a laser (Benthall, J., *Science and Technology in Art Today* (New York: Praeger, 1972), illustration p. 87, reprinted courtesy of M. Benyon).

Kip Siegel, successively a Willow Run Division Head, University of Michigan professor and Ann Arbor entrepreneur, tempted audiences with 'what you could do with true 3-D':

> If you went to a classroom, and were taking pictures of your children—say, playing in kindergarten, and one of the children ran behind the other children—instead of stopping taking movies, you continue taking movies, and then when you get home and you show those movies of your children, when your child has ducked behind other children, all you need to do is move your head, and you can follow your child, as your child is running behind other people . . . There are other properties associated with holograms that you'll be able to see . . . You can take off half the picture—you can cut off half the picture—and you still have the ability to see the whole picture.[45]

Such weird properties of Leith–Upatnieks holograms (but not all of their successors) captivated imagination and spawned enduring popular interest. Hinting both at magicians' tricks and reassuringly familiar technologies, they also suggested a cultural place, and a social future, for holograms.[46]

[46] Some of these notions were overgeneralized for later audiences. A featureless surface, and a part that reproduces the whole, are not properties of holograms that reconstruct images near the plate itself. These so-called image-plane holograms (Section 6.4) can still be understood as windows, though: only portions of a window permit a view of an object pressed up against it. Generally speaking, the closer that a scene is to the window, the more localized its viewing will become and the closer it will behave like a photograph. Arthur C. Clark's remark that 'Any sufficiently advanced technology is indistinguishable from magic' can be recast: our assessment of the magic of holograms is the result of the analogies that we unconsciously adopt, or that others frame for us (Clarke, A. C., 'Hazards of prophecy: the failure of imagination', in: A. C. Clarke (ed.), *Profiles of the Future: An Enquiry into the Limits of the Possible* (London: The Scientific Book Club, 1962), p. 36).

[45] Siegel, K. M., 'Speech to Conductron Missouri employees' (C. Charnetski collection, 1966).

6
Holograms and Progress

Holograms illustrated the power of the new postwar style of muscled scientific research, and were well-placed to visualize the onward march of the modern world. Several ingredients combined to represent holograms as the future: fascinated technologists exploring a wholly new field; optimistic promoters accustomed to incremental improvements, and overwhelmed by new and unexplored opportunities; a receptive public primed for novel media experiences; and, more generally, a zeitgeist confident of the inevitability of techno-scientific progress.

By the mid-1960s, wider publics were being tempted to speculate about the implications for the near future. While opportunities to experience holograms remained limited, accounts multiplied in popular media. Mass-mediated forecasts about the progress of holograms—unconstrained by reality—suffused modern culture. And, argued Umberto Eco, holograms would become particularly popular in America,

> where, if a reconstruction is to be credible, it must be absolutely iconic, a perfect likeness, a 'real' copy of the reality being represented.[1]

As the latest scientific innovation spring-boarding from other recent inventions, holograms assured consumer wonders in the rosy long-term.

6.1 REPRESENTING MODERNITY

Holograms tapped into cultural understandings that had been developing over the previous century. The industrial age impinged increasingly on every detail of life. As traced in the previous chapters, the machine-made world was both seductive and unavoidable. Inventive novelty and increasing dependence on the products of technology had become the hallmarks of modernity.

The trigger of wartime development showed convincingly that research paid dividends, and the postwar period amplified these cultural impressions. Wartime innovation led directly to postwar classified research, but new wonders surfaced with encouraging

[1] Eco, U. and W. Weaver, *Travels in Hyper-Reality: Essays* (London: Picador, 1986), p. 1.

regularity. The link between innovation and social benefits began to seem obvious. In medicine, the wartime development of penicillin had led to new families of antibiotics, and anti-psychotic medications began to substitute for confinement in mental hospitals. The Manhattan Project had invented not merely nuclear weapons, but the burgeoning new fields of nuclear power generation and nuclear medicine. Wartime calculators were succeeded by postwar computers for accounting, scientific calculation and, soon, the real-time demands of command-and-control systems. Motivated by military needs, electronics was being revitalized by transistors and integrated circuits. Advances in rocketry now promised to realize interwar science fiction fantasies. And an expanding petrochemical industry concocted a spectrum of synthetics to build, clothe and fashion the postwar world.[2]

As detailed in Chapter 5, holograms, too, emerged from such covert postwar environments to impress the public. Holograms were especially potent because they combined distinct threads of modernity. And they were the products of lasers, which were themselves brand-new.[3]

Lasers provided their own set of magic tricks: their beams remained pencil-thin over incredible distances, and in special cases, they could perform special tricks like transmitting telephone conversations, selectively burst coloured balloons or pierce razor blades. 'Modern optics', as it was dubbed by its practitioners, had little connection with traditional and mundane applications such as cameras, telescopes or even 3-D movies. Exotic rather than quotidian, it was brimming with possibilities.

Holograms and lasers were products of a new kind of scientific laboratory, no longer inhabited by the lone mad scientists, electrical machines and jumping sparks common to 1930s movies.[4] Media representations of postwar labs quickly became standardized. Magazine photographs and television news items depicted tie-wearing technologists quietly at work on government and corporate projects. Even in the restricted space of the lab, those scenes often included two or more collaborators. The Willow Run engineers, more habituated to open necks and shirtsleeves, conformed to media expectations.

Clean optics and precision steel surfaces characterized these well-funded environments; the frugal 'string and sealing wax' approach familiar to interwar physicists was banished. Bright red laser beams and white-coated scientists were ideal for colour television, which was becoming more common in the USA and introduced in Britain and other European countries in 1967. For the same reasons, stories on holography could enliven colour picture magazines of the 1960s (Figure 6.1). In a special 1966 issue, for

[2] See, for example, Bud, R., *The Uses of Life: A History of Biotechnology* (Cambridge: Cambridge University Press, 1993); Agar, J., *The Government Machine: A Revolutionary History of the Computer* (Cambridge, MA: MIT Press, 2003); Johnston, S. F., *The Neutron's Children: Nuclear Engineers and the Shaping of Identity* (Oxford: Oxford University Press, 2012); McDougall, W. A., *The Heavens and the Earth: A Political History of the Space Age* (New York: Basic Books, 1985); Meikle, J. L., *Plastic: A Cultural History* (New Brunswick, NJ: Rutgers University Press, 1997).

[3] Gilmore, C. P., 'How lasers are going to work for you', *Popular Science* (1970): 48–53.

[4] Frayling, C., *Mad, Bad or Dangerous? The Scientist and the Cinema* (London: Reaktion, 2005).

THE TECHNOLOGY OF TOMORROW?

Hologram is made *by using laser beams to illuminate both subject and reference source. Scenes were shown to editors during Mr. Hunt's address.*

Fig. 6.1 The technology of tomorrow. Bell Telephone Laboratories publicity photo (Hunt, G. Carlton, 'The impact of visual communications on industry', *Business Screen* 28 (1967): 24–5).

example, *Life* magazine described 'the latest far-out form of photography' with an item on Leith and Upatnieks shot by Fritz Goro, a photographer who specialized in portraying the excitement of modern science via stunning visuals.

Descriptions, too, became formulaic. These dedicated 'scientists' (more often than 'engineers') worked intensively. Their research illustrated the work ethic and corporate motto of IBM, *THINK*. *Life* magazine described not just magic, but hard work: a single hologram 'required three beam splitters, five lenses, thirteen mirrors and a laser—and took more than a week to set up'. Holograms epitomized the mixing of technology, professional dedication and goal-driven research.[5]

6.2 EXPRESSING NATIONAL STATUS

Such indicators of progress and modernity had political dimensions, too. As the last war had hammered home, scientific research and development had national benefits. Technological successes signalled expanding economies. This message was trumpeted not only

[5] Steinman, M., 'Tools to reach beyond the bounds of man's vision', *Life*, 61 (26), 23 Dec 1966: 104–19; quotations p.116. On Goro (1901–86), see Smith, C. Zoe, 'Fritz Goro: émigré photojournalist', *American Journalism* 3 (1986): 206–21.

in America to explain the benefits of the free market, but also in the Soviet Union, where Nikita Khrushchev heralded the birth of a second 'scientific-technical revolution', and in Britain, where Harold Wilson's government sought to marshal 'the white heat of technology'.[6] These descriptions seemed excellent fits to the achievements of the 1950s 'atomic age' and the just-beginning 'space age' of the 1960s.

As a product of what the outgoing President, Dwight D. Eisenhower, had called the 'military-industrial complex' in 1961, the hologram embodied these themes.[7] The University of Michigan research was an unexpected spin-off of Willow Run defence contracts, and was rapidly explored and developed by companies participating in contract research. A rival of the Willow Run teams, Prof George Stroke publicized his own work by situating it in this global political context. His own work on holograms viewable in 'white' light was, he said, 'a hair-thin balance between sophisticated mathematics and extremely refined experimentation [. . .], new miracles comparable to a successful Apollo moon shot in this field'. When Yugoslavian immigrant Stroke hinted at his patriotism by displaying a hologram of Abraham Lincoln for reporters, his foe, Midwesterner Emmett Leith, artlessly countered with a large hologram of an army tank. In a similar vein, Stroke promoted his creation of colour holograms with an example showing soldiers in front of the red, white and blue American flag.[8]

In the context of the times, narratives of government and corporate support often prevailed. As early as 1967, the *New York Times* introduced holograms to readers of its Business & Finance section as a likely investment opportunity. Holograms encouraged forecasts of 'wide military applications', improved radar systems and engineering study of deformations but—most importantly of all—potential profits in the generous funding environment of the military-industrial complex.[9] More broadly, the corporate origin of most American holograms channelled a subliminal ideological dimension via their images, to which cultural critic Mark Fisher has applied the term 'capitalist realism'.[10]

But technologies to demonstrate patriotism and national prestige were not limited just to the USA. Although Yuri Denisyuk's research at the beginning of the decade was little appreciated by his contemporaries until the American developments; by the late 1960s he and his peers began to achieve national recognition. Denisyuk was awarded the Lenin Prize in 1970, and Pyotr Kapitsa, a Nobel Prize winner and powerful member

[6] Hoffman, E. P., 'Soviet views of the "Scientific-Technical Revolution"', *World Politics* 30 (1978): 615–44; Wilson, H., 'Conference speech, Labour Party Annual Conference', Scarborough, 1 Oct 1963.

[7] Eisenhower, D. D., 'Farewell address to the nation', 17 Jan 1961.

[8] Stroke, G. W., 'Breakthrough in "lensless photography" sets stage for 3-dimensional home color television and a possible multi-million industrial explosion in electro-optics', Press release from Univ. of Michigan for 14–18 March 1966 OSA meeting.

[9] Sullivan, W., 'New photo technique projects a world of 3-dimension views', *New York Times*, 19 Mar 1967, 1.

[10] Fisher, M., *Capitalist Realism: Is There No Alternative?* (Ropley, UK: Zero Books, 2009). 'Capitalist realism' was originally a German pop art movement of the early 1960s. Unlike this original ironic usage, the term later described features of commercial art in western societies, but used by Fisher in the broader sense of a perceptual framework for aesthetics.

of the Soviet Academy of Sciences, ensured that his reputation was rehabilitated. Denisyuk holograms could accurately be touted as an important variant that pre-dated the Willow Run achievements—a priority claim that contributed to Soviet national status for audiences at home and abroad. And, just as American labs displayed holograms that illustrated American values, Soviet holograms were created of nationally meaningful subjects. Among the first products of the holography lab of the Institute for Optics and Spectroscopy in East Germany were holograms of busts of Karl Marx and Vladimir Ilyich Lenin.

Soviet holograms responded to national tastes and contexts more subtly, too. Ukrainian holographers conceived a role for holograms as proxies for museum exhibits. Recording precious historical objects, these national treasures were exhibited in travelling exhibits by bus and train. When displayed internationally, these holograms simultaneously showed off Soviet high technology and cultural heritage.[11]

In the more diffident political contexts of the UK and other western European countries, there were fewer overt references to national histories or shared ideologies. For Dennis Gabor and the most important of his early co-investigators, Gordon Rogers (1916–2006), their experiences seemed, to their chagrin, to exemplify the hackneyed narrative of British innovation followed by American business success.[12]

6.3 ANTICIPATING CONSUMER WONDERS

The earliest popular understandings, though, fitted a narrative of consumer benefits. The century-old appeal of three-dimensional imagery had been sustained by stereoscopes, lithographed colour stereo views, 3-D movies and children's anaglyphic comic books. Pre-war audiences had expected television to rapidly extend to the third dimension (Figure 3.8), and their postwar counterparts were seduced by similar media extrapolations (Figure 6.2).

The pre-eminence of television as the latest consumer medium suggested that, while they started as novel photographs, holograms would inevitably underpin the television of the future. Emmett Leith himself—providing hints of the classified links between the electronics of side-looking radar and the optical processing of its images—suggested to reporters the potential of holographic television. At the 1965 National Electronics Conference in Chicago, he and Upatnieks showed more holograms to a much wider audience.

[11] *Holography in the USSR: the Art and Science of the Soviet Union exhibition catalogue* (York, UK: 1981); Markov, V. B. and G. I. Mironyuk, 'Holography in museums of the Ukraine', presented at Three-Dimensional Holography: Science, Culture, Education, Kiev, USSR, 1989; Morgan, A., 'Treasures trapped in light + exhibit of Ukrainian holograms in York', *History Today* 39 (1989): 5. See also Section 9.3.

[12] In 1956, Rogers had written 'I see relatively little future for it, and look forward to doing something else', but a decade later, he complained to Gabor that their work had contained most of the elements later employed by the Americans (Rogers, G. L. to A. V. Baez, letter, 19 Jul 1956, Sci Mus ROGRS 6; Gabor, D. to G. L. Rogers, letter, 5 Aug 1966, Sci Mus MS 1014/15).

Fig. 6.2 Forecast 3-D technologies. Arthur Radebaugh (1906–74), an illustrator and advertising artist, worked principally in the American Midwest car industry. His cartoons were nationally syndicated in newspapers 1958–61, reaching circulations of some nineteen million readers (Radebaugh, A., *Closer Than We Think*, 21 Dec 1958, with permission of Tribune Content Agency / Tribune News Service Reprints).

Leith noted that 'applications that come to mind immediately are television and motion pictures' and that it was possible, in principle, 'to produce a hologram television system. However, today's television equipment is not adequate and each receiver would have to contain a laser light'.[13] To another reporter, Leith cautioned, 'that is really a wild use because even though it is technically possible, the difficulties are great'.[14] The fundamental problem, Leith appreciated, was that the amount of information to be transmitted far exceeded any practical communications bandwidth.

With the same technical insights but a better eye to publicity, his nemesis, George Stroke, wrote his own press releases a year later touting the potential of holographic television.[15] Anonymous sources in the electronics industry even hinted that a prototype holographic TV system was soon to be unveiled, with predictions of a consumer market within the decade. To oppose such forecasts seemed narrow-minded and even elitist; one commentator suggested that naysayers like Leith and Dennis Gabor 'would rather see it bloom as a scientific field, not for such entertainment as the broadcast of a "Punch and Judy" show'.[16]

Reporters beyond America channelled similar claims. A science correspondent for the British national newspaper *The Observer* forecast that holograms 'could lead to 3-D colour

[13] 'National Electronics Conference, Chicago', *Pontiac Press*, 28 Oct 1965, 33.

[14] Veteto, B., 'Laser may make 3-D TV possible', *Dallas Times Herald*, 31 Mar 1965: A-31.

[15] Stroke, G. W., 'Breakthrough in "lensless photography" sets stage for 3-dimensional home color television and a possible multi-million industrial explosion in electro-optics', Press release from Univ. of Michigan for 14–18 March 1966 OSA meeting.

[16] 'Holographic TV: picture is bright', *Electronics*, 28 Nov 1966: 25.

television of uncanny realism, to computers that can read, to 3-D portraits that you can hang in front of a window, to unbreakable codes for spies'.[17]

But holograms proved particularly effective as marketing propaganda in the USA. Keeve M. (Kip) Siegel (1923–75), the University of Michigan professor who had left Willow Run Labs to found Conductron Corporation, adopted a strategy of 'selling the promise of technology to investors'.[18] This ploy relied on the creation of innovative demonstration projects to impress stockholders and the press. Holograms—straddling militarily-funded expertise and corporate innovation—were ideal for this. In 1966, addressing the 1400 employees of his latest acquisition, he focused on the wonder encapsulated in the hologram and the inevitability of its commercial exploitation:

> We are doing major work in three-dimensional television and towards three-dimensional home movies . . . I am hoping that by the year 1976 that the United States will have, as far as new products are concerned, only three-dimensional television and three-dimensional movies on the market. I would not expect two-dimensional processing, two-dimensional television, two-dimensional home movies to continue—that's my personal belief. I don't think people will buy things that are antiquated. If that volume comes to pass, then I think that the price will be equivalent to the price of two-dimensional television today, assuming that the dollar does not become too inflated. In any real measure of the absolute buying power of the absolute dollar, I think you could buy a 3D set for the price you would buy a 2D set today, and I'd give similar answers on processing and similar prices on film.[19]

Such seduction underpinned the sale of Conductron to the aircraft manufacturer, McDonnell-Douglas (M-D) that year. The seduction of potential investors guided much of Siegel's research and development on holograms. Conductron engineers developed a way of creating a short holographic movie to win a bet with Jim McDonnell, co-owner of M-D. The demonstration was effective, but was limited to a few dozen holograms recorded one by one and then animated to create repetitive, jerky motions of a few seconds duration that could be viewed by a handful of observers.[20]

Moving from such restricted demonstrations to more promising markets required even bolder claims. One of Siegel's more outlandish predictions was his stated intention to 'holograph' the next Olympic Games. The task conjured up ancillary magic: it would require intense pulsed lasers—impossibly powerful and dangerously bright, in fact. The company consequently developed the first pulsed laser suitable for holograms in 1967, but used it tentatively for 'flash' holography of a dog, a hand, and an engineer's face.[21] Over the next six years, Conductron turned from investors to potential customers, trying to tempt advertisers, artists and affluent buyers to use the

[17] Davy, J., '3-D pictures in the round', *Observer*, 15 May 1966, 13.

[18] Gillespie, D. to SFJ, interview, 29 Aug and 4–6 Sep 2003, Ann Arbor, MI, SFJ collection.

[19] Siegel, K. M., 'Speech to Conductron Missouri employees' (C. Charnetski collection, 1966).

[20] Charnetski, C. to SFJ, interview, 3 Sep 2003, Ann Arbor, MI, SFJ collection; Cochran, G. D. to SFJ, interview, 6 and 8 Sep 2003, Ann Arbor, MI, SFJ collection.

[21] Siebert, L. D., 'Large-scene front-lighted hologram of a human subject', *Proceedings of the IEEE* 56 (1968): 1242–3.

system.[22] As Gary Cochran, the manager responsible for the work recalled, their environment was 'an imaginary Disneyworld of scientists'.[23]

Despite their recognized impossibility within the company, Siegel's predictions carried the ring of insider authority. Like falling dominoes, prediction triggered prediction in an unstoppable cascade. The same theme was echoed that year in the advertising industry:

> Some day at a meeting of the future, without special red-green glasses, you'll really see living three-dimensions in a motion picture or closed-circuit television presentation. You'll be able to get out of your chair to peek around the figure on the foreground of the screen to see what's behind it.[24]

The following year, a typical article promised that the house of the near-future would have a TV screen that was 'three-dimensional or holographic, enabling you to look around corners almost as though you were inside the scene being projected'.[25] The President of the Society of Motion Picture and Television Engineers argued that the consumer experiences of previous decades were a good indicator of coming developments. Citing the growth of amateur photography and cinematography, wide-screen 'and even 3-D' movies, and the recent 'colour explosion' in photography, movies and television, his address culminated in the latest rumours of miraculous holograms:

> Several companies are said to be working on three-dimensional movies in which a viewer can look at a screen from any angle and see the object depicted as if it were seen from that angle. One Cleveland based corporation has forecast that three-dimensional color TV will be with us by 1971 . . . [26]

And, right on schedule, RCA Corporation introduced SelectaVision, a video-playback system based on holographically recorded tapes. While the technology provided only a good quality (two-dimensional) recording of a conventional television picture, its promotion embedded holograms further into consumer consciousness. A *Life* magazine article channelled the inventors' hype:

> How about an 8-foot John Wayne in 3-D?
> [. . .]RCA's prospects look even brighter for the not-so-distant time when TV sets have grown into wall screens and 3-D is a reality. Fully three-dimensional still colour photographs have already been produced using the hologram process. Within 15 years it may be possible to apply the same technique to moving pictures—but only using RCA's system, or one like it. Then, as one expert suggests, viewers may gaze, in their own living room, upon an eight-foot-tall image of John Wayne (presumably still going strong) and even walk around him.[27]

22 Siebert, L. to SFJ, interview, 4 Sep 2003, Ann Arbor, MI, SFJ collection.

23 Cochran, G. D. to SFJ, interview, 6 and 8 Sep 2003, Ann Arbor, MI, SFJ collection.

24 'Laser beam creates new A-V medium', *Sales Meetings*, 15 May 1966: 52–7.

25 Ward, B., 'House of tomorrow', *Radio-Electronics* (1967): 21–3.

26 Hunt, G. C., 'The impact of visual communications on industry', *Business Screen* 28 (1967): 24–5.

27 Groskinski, H., 'Cassette TV: the good revolution', *Life*, 69 (16), 16 Oct 1970: 46–53, quotation p. 51. The same tropes circulated at the end of the decade regarding the advertising potential of holograms: 'The chance to project a fifty foot bottle of Guinness over Piccadilly Circus is too good to pass up!' (Austin, D., 'Exhibition: Light Fantastic 2', *New Scientist*, 19 Jan 1978: 171).

Multi-media creator Jim Sant'Andrea predicted, 'Videotape and holography will be used not only on screen, but for other projections as well'.[28] Joseph Strick (1923–2010), a technology entrepreneur and film director, publicized his intention in 1970 to make a 'hologram motion picture, shot with the use of laser rays', in which 'actors will be seen in the round, and will even jump into the laps of the audience, without anyone wearing trick glasses'. Supported by impressive credentials as the founder of a series of technology businesses during the late 1950s and with a reputation as an American new wave cinema pioneer, Strick's ambitious claims were reported straight-faced:

> Holograms can be projected so that the audience can see completely around the image presented—and a hologram actor can even mingle with the audience, like an incorporeal ghost . . . Strick says work is already under way on a projection system that can be seen by 400 to 600 people . . . His first hologram picture should be ready next year, a movie-in-the-round that may make the Woodstock generation feel slightly threatened by the pace of technological change.[29]

Needless to say, the promises never materialized.

The founder of Conductron, which had provided the template for hologram hype, continued to pursue its investment potential. After selling the company to McDonnell-Douglas in 1966, Kip Siegel left to found a new company, KMS Industries, where he again used holograms as the bait for further venture capital. Bringing some of his senior engineers with him, Siegel proselytized and promoted to new audiences. As Gary Cochran remembered, Siegel 'was talking about three dimensional TV, and regardless of the bandwidth I said it needed, he said "Aw, it will come along someday". I remember listening to him give the pitch to people . . . It sounded wonderful, but I had no idea how to do it technically'.[30] When the effectiveness of the hologram pitch began to wane by 1969, Siegel founded yet another company, KMS Fusion, to pursue the promise of nuclear fusion power generation.

6.4 ENGINEERING THE FUTURE

While seeding consumer optimism, companies were also investigating more achievable goals. Conductron and a handful of other Ann Arbor companies were manufacturing holograms in large numbers as advertising demonstrations, trade-show give-aways and magazine inserts, but discovered that markets, and even the necessary background awareness, were still very limited.

Mechanical engineers of the late 1960s were more eagerly exploring a new field—holographic stress analysis—which allowed the mapping of miniscule deformations. When mechanically or thermally distorted, test objects would reveal holograms bathed

[28] Sant'Andrea, J., 'Messages of a multi-media maker', *Business Screen* 32 (1971): 13–6.

[29] 'Strick will make laser film', *Los Angeles Times*, 21 Aug 1970, 14.

[30] Cochran, G. D. to SFJ, interview, 6 and 8 Sep 2003, Ann Arbor, MI, SFJ collection.

in fringe patterns, from which movements could be analysed. The technique carried the potential to transform design and testing, especially for high-precision industrial and military equipment.

Often supported by government contracts from NASA, the Advanced Research Projects Agency (ARPA), the Office of Naval Research (ONR), the Army Research Office (ARO) and American Air Force, companies were exploring new types and uses for holograms. During the late 1960s, numerous threads were pursued.[31]

Varieties and capabilities of holograms kept multiplying through the decade, with 'white light' holograms gaining the widest attention. The need to use a laser to illuminate the hologram was a severe drawback even for well-funded corporate labs. Leith and Upatnieks had borrowed their first laser from another Willow Run group in 1961, and the expensive and temperamental devices were still uncommon by the mid-1960s. The research of Yuri Denisyuk in Russia had invented a different variety of hologram that combined some of the characteristics of Leith's and Upatnieks' type with the turn-of-the-century ideas of Gabriel Lippmann (Section 2.4). The method could reconstruct a holographic image even when lit by a regular 'white' lamp (although the lamp had to be small—nearly a point-source—for best results). When Denisyuk was able to refine his technique by recording his holograms with a laser, his 'white-light reflection' holograms proved impressive and liberated audiences from the need to use a laser (Figure 6.3; Box 6.1).

These achievements remained little known in Western countries during the 1960s, however. Denisyuk's papers were terse and poorly translated, and his status at home was obscure. Similar results were rediscovered, published in English-language scientific papers and explored by American commercial firms in 1967.[32]

One reason for this overlap and confusion of knowledge was the explosion of international research, nearly all of it exploratory and opportunistic, and only weakly tied to underlying analyses—exactly the inverse of the 1950s theoretical work on holograms. Adding to the intellectual competition was the promise of patents and confidence in inevitable profits.

Confusing matters further was the variety of effects that could be produced by minor variations in how holograms were made. This was fertile magic. 'White light' holograms—viewable with regular light sources—could be made either by Denisyuk's elegant method, or by the simple expedient of focusing the image in the plane of the hologram plate with an extra lens. These holograms, too, generated sharp images when viewed in white light, but had their own magical properties: the images straddled the plate itself. Such 'image-plane' holograms were not just three-dimensional: they were literally within grasp and yet immaterial (Box 6.2). During 1968, the characteristics of these new variants enthused their creators and the audiences privileged to witness them.[33]

[31] See also Johnston, S. F., *Holographic Visions: A History of New Science* (Oxford: Oxford University Press, 2006), Chap. 6.

[32] See Johnston, S. F., *Holographic Visions: A History of New Science* (Oxford: Oxford University Press, 2006), Section 7.3.

[33] See Johnston, S. F., *Holographic Visions: A History of New Science* (Oxford: Oxford University Press, 2006), Section 7.4.

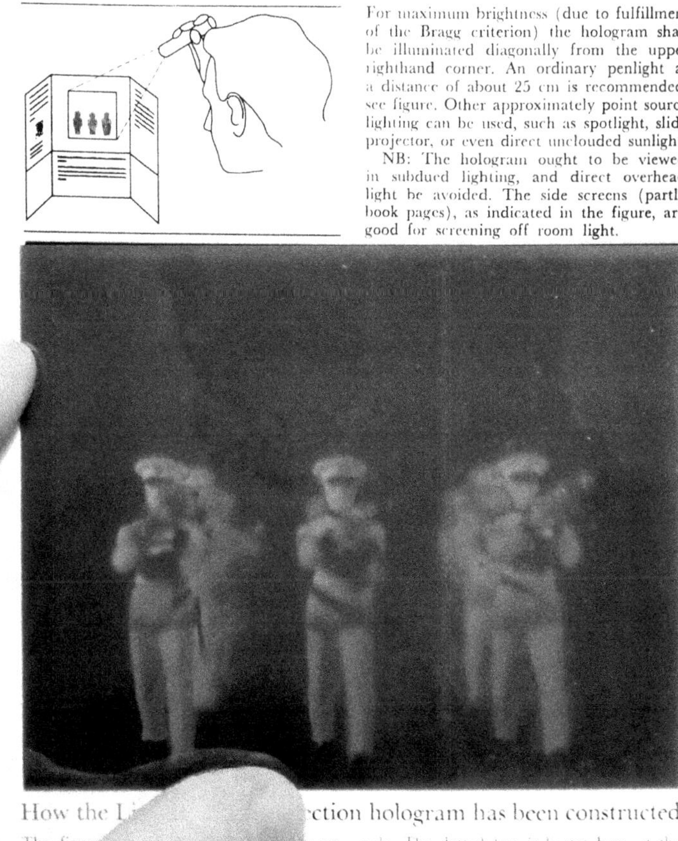
How to view the Lippmann type reflection hologram

For maximum brightness (due to fulfillment of the Bragg criterion) the hologram shall be illuminated diagonally from the upper righthand corner. An ordinary penlight at a distance of about 25 cm is recommended, see figure. Other approximately point source lighting can be used, such as spotlight, slide projector, or even direct unclouded sunlight.

NB: The hologram ought to be viewed in subdued lighting, and direct overhead light be avoided. The side screens (partly book pages), as indicated in the figure, are good for screening off room light.

Fig. 6.3 Image-plane reflection hologram (Gabor, D., 'Holography, 1948–1971', in: *Les Prix Nobel en 1971* (Stockholm: The Nobel Foundation, 1972), pp 169–201; insert pp 176–7; hologram produced by Conductron Inc; S. Johnston photo).

Research was also directed at the difficult goal of holographic television. Engineers recognized that this would require a severe trimming of the amount of information in the transmitted holograms. One line was the work of Stephen Benton at Polaroid, who developed a simple method of reducing the information content by tenfold or even a hundredfold. His technique sacrificed vertical parallax: observers would see all the magical qualities of a Leith–Upatnieks hologram—including looking *around* a foreground object to see one behind it—but they would be unable to look *over* or look *under* it. When viewed

Box 6.1 Viewing Denisyuk holograms

When lit by a laser or white-light source, a Denisyuk hologram (also known as a 'reflection' or 'Lippmann' or 'Bragg–Lippmann' hologram) reveals a three-dimensional image behind the plate. The experience is exactly like looking at the original scene through a window. The image is life-sized and usually less than a metre or so deep.

Unless the white-light source is point-like (e.g. from a small-filament incandescent bulb), the image will be slightly fuzzy, especially more distant from the hologram surface.

The orientation and spread of the light source must match the arrangement that was used when originally recording the hologram to avoid distorting this virtual image.

Denisyuk holograms are capable of reproducing full-colour three-dimensional images if suitably recorded with three or more laser wavelengths. More typically, when viewed with a white-light source the image is single-coloured (monochromatic), often in green-yellow tones.

With suitable recording arrangements, a reflection hologram may also reveal an image that passes *through* the plate or hangs in front of it. This variant is known as an 'image plane' or 'H1 / H2' or 'second generation' reflection hologram.

Box 6.2 Viewing image-plane holograms

Holograms of all varieties can be created to reveal images either behind, in front of, or actually straddling the hologram plane. Those that straddle the hologram are known as 'image plane' or 'H1 / H2' or 'second generation' holograms.[34] The experience is uncanny because part of the image appears to hang in space, seemingly projected from the hologram surface.

In most respects, the images appear like their 'behind the plate' counterparts. The illusion of reality is slightly compromised, though, because the viewing frame or window is no longer the physical edge of the hologram itself. Instead, the viewer may detect an invisible 'port-hole' or aperture that vignettes the field of view.[35] As the viewer moves off-centre, the image more or less suddenly fades away.

[34] The image can be made to straddle the plate by either using a lens to focus the real image of an object (a 'one-step' image plane) or by using the real image from a first master hologram, H1, to create a second, H2 (known by hologram producers—but seldom viewers—as an 'H1 / H2' or 'second generation' hologram).

[35] Smith, C. W., '3-D or not 3-D', *New Scientist* 102 (1984): 40–4.

Box 6.3 Viewing rainbow holograms

A rainbow hologram (also known as a 'Benton' or 'white-light transmission' hologram) is lit by a point-like or line-like white-light source, and reveals a characteristic three-dimensional image. The experience is like looking through a window, but with two peculiar characteristics. First, the colour of the image depends on the viewer's vertical position: moving up or down will shift the colour, as if viewing through a rainbow. Second, this vertical movement does not show parallax: it is impossible to look over or under nearby objects in the scene.

When lit by a laser, the image appears to be a horizontal letter-box slit, through which the viewer can peer to see a wide horizontal view.

Rainbow holograms may be transparent—to be looked through—or made reflective by metallizing the surface. Reflective rainbow holograms have been particularly popular for reproduction in magazines, stickers and cards.

As with other varieties of hologram, a rainbow hologram may, if suitably recorded, also reveal an image that passes *through* the plate or hangs in front of it. This 'image plane' variant is, in fact, the most commonly reproduced type of rainbow hologram.

in laser light, the reason for this becomes obvious: it is like peering through a long, narrow letter-box slot at the hologram scene, allowing a full horizontal view but only a fixed vertical position. When viewed in white light, though, the result was a new variety of magic. Benton's innovation, popularly known as the 'rainbow hologram', revealed a full view of the original image, but tinted in a vertical spectrum of colours. As the observer bobbed up or down, no hidden details came into view, but the overall colour could range from deep red to bright yellow to dim blue (Box 6.3).[36]

At the end of the 1960s, then, holograms were rapidly diversifying but still reaching limited audiences. That first wave of innovation had invigorated corporate engineers, media professionals and entrepreneurs. Wider publics, though, heard belatedly and little about holograms, and relatively few, as yet, were fortunate enough to see them.

[36] See Johnston, S. F., *Holographic Visions: A History of New Science* (Oxford: Oxford University Press, 2006), Section 7.7.

PART C
Hologram Cultures

As holograms became more accessible, wider publics appropriated them for new uses. Enthusiasts, like predecessors earlier in the century, mastered and extended the technology in new directions. Hologram-making became a pastime, an art, a medium of education and an expression of philosophical ideals. Seduced by promoters, consumers aspired to hologram products that would transform daily life. So compelling were the dreams that enthusiasts carried them into the future through fantasy worlds.

7

Holograms for Enthusiasts

Our little laser group, we were an ear to the lonely voices out there, to the people who were doing it on their own—those folks who were in their caves with no one nearby to relate to. And it was great, because everybody was connecting; a field full of intelligent, creative, manually dexterous people . . . all from the uniqueness of holograms.[1]

Vince DiBiase, Santa Clara, 2003

During the late 1960s, only the most dedicated amateurs were able to see, and very occasionally make, their own holograms. This first wave of public engagement mirrored cultural experiences with earlier technologies. The enthusiasts who made their own holograms were the latest generation of technical amateurs. These precursor cultural contexts defined what it meant to be a technical enthusiast and amateur scientist, and provided models for how holograms would become accessible to wider publics. But hologram enthusiasts were also unlike their predecessors: they entered a brand-new field that had little cultural visibility and no established commercial applications.

At the turn of the twentieth century, photographic hobbyists had served as intermediaries between professional photographers and wider audiences. According to the popular writings of the day, tinkering could lead to inventing, which might be linked to status, wealth and even adventure. Four decades before holograms, nascent radio technology had attracted broad audiences. Appropriated by amateurs who had picked up the technology during the First World War and developed in fresh directions, it was also adopted by listeners seeking modern entertainment at home. At the same time, telescope-making was being popularized as a hobby by a network of interwar publishers. The Second World War had consolidated this culture of scientific amateurs, with science clubs sprouting up to support national competitiveness.

But the pinnacle of amateurism was reached with the postwar 'baby boom' generation. The war's end fuelled technical enthusiasms by making available war-surplus components—optics, electronics and mechanical devices; innovation percolated down quickly to amateurs in ham radio; and Cold War competition encouraged public anxieties and a national urgency to produce competent technologists.

[1] DiBiase, V. to SFJ, interview, 21 Jan 2003, Santa Clara, CA, SFJ collection.

As with making photographs from the 1890s, experimenting with radio from the 1920s, building telescopes during the 1930s and hobby rocketry in the 1950s, amateur explorations of holograms developed in a cultural environment in which technical pastimes, old and new, were imbued with scientific zeal. Cultural associations stressed individuality, autonomy, technical competence, thrill of discovery and even national competitiveness. From the mid-1960s, hologram hobbyists became the latest participants in a material culture that had been growing among niche audiences since the beginning of the century. These special interest communities had no single name; instead, they were islands of enthusiasm for particular technologies and, in some cases, the arcane sciences that underpinned them.

7.1 HOBBYISTS, TINKERERS AND TECHNICAL ENTHUSIASTS

Hologram enthusiasms emerged from earlier technical pastimes that had flourished during the twentieth century. The zeal to collect objects, for example, is widespread, particularly for children. Collecting coins, stamps, dolls or bubble-gum cards marks out a specialist activity that can be pursued intensively into adulthood and even mark out status amongst peers. Alternatively, traditional hobbies could focus on craftwork. These activities generally involved gendered manual skills such as embroidery, gardening, woodworking or model-making, where again personal achievement could be compared with others. Both aspects of hobbies—collecting and making—could provide a satisfying leisure activity, but also foster a sense of personal proficiency and group identity.[2]

But from the beginning of the twentieth century, these traditional attractions were bolstered by pastimes focused on technologies. For three successive generations between the 1890s and 1960s, new inventions transformed life and aspirations. As explored in Chapter 2, photography melded chemistry, physics and aesthetics. Electrical technologies for lighting, communication and mechanical power began to invade public spaces, institutions and some middle-class homes.[3] Gasoline engines became ubiquitous in urban and rural environments,

[2] Gelber, S. M., *Hobbies: Leisure and the Culture of Work in America* (New York: Columbia University Press, 1999); McKibbin, R., 'Work and hobbies in Britain, 1880–1950', in: J. Lerner (ed.), *The Working Class in Modern British History: Essays in Honour of Henry Pelling* (Cambridge: Cambridge University Press, 1983), pp 143–5; Havighurst, R. J. and K. Feigenbaum, 'Leisure and lifestyle', *American Journal of Sociology* 64 (1959): 396–404; a related example is Moorhouse, H. F., 'The work ethic and leisure activity: the hot rod in post-war America', in: P. Joyce (ed.), *The Historical Meaning of Work* (New York: Cambridge University Press, 1987), pp 257–81. Such cultural studies have highlighted the relationship between work, leisure and capitalist society, and argue that the taking up of hobbies during the late nineteenth century, especially among the working class, was both a reflection and a limited resistance to industrialization and the free market. I argue here that some technical hobbies—and particularly holography—followed an arc that took them from being a mirror of workplace values to a rejection of them. The peak occurred during the 1960s, when dominant culture became infused with the ideals of science, and a counterculture developed to critique those ideals.

[3] Hughes, T. P., *Networks of Power: Electrification in Western Society, 1880–1930* (Baltimore: Johns Hopkins University Press, 1983).

and skills in repairing automobiles and farm equipment often led to the acquisition of other technical competences and interests. And steam power, a familiar backdrop from the early nineteenth century, ironically became a symbol of modernity and the Machine Age as it was declining between the World Wars.[4]

New pastimes incorporated the traditional attractions of collecting and producing, but placed an exaggerated emphasis on the technical background and satisfactions of expertise itself.[5] Their mystery added attraction: how did a car engine work, for example, and what exactly was electricity? On the other, they fostered a growing culture of tinkerers, offering empowerment for those who mastered the new technologies. These intellectual poles, dividing the inept adopter from the adept devotee, nevertheless united modern generations in terms of the technologies that they both embraced. And by entraining wider audiences in their wake, the early adopters and technical enthusiasts shifted cultural values inexorably towards novel technologies. These new capabilities could intrigue, inspire and even excite.

Such activities could also invigorate otherwise mundane lives. Members of unprivileged niches of society could exert their individuality, achieve esteem among peers, and even augment income by their proficiency in hobbyist domains. Certain hobbies could differentiate individuals from their neighbours and provide routes to avocations, if not careers.

This expanding cultural space was inhabited by an evolving constellation of special interests which provided a framework for twentieth-century amateur culture and its expression in holography. Rehabilitating and situating these earlier technical enthusiasms is a necessary prelude to understanding the role of holograms in late twentieth-century culture.

7.2 GROWING AMATEUR SCIENTISTS

Books and magazines provided templates for modern technical enthusiasms, conditioning the professionals and enthusiasts alike who eventually took up holography. From the turn of the century, and especially in the USA, a new genre of popular fiction had preached the values of rational deduction, inventiveness and the potential of technologies for generating adventure. A series of periodicals aimed these enthusiasms and latest inventions at the working- and middle-classes. The result was an infrastructure of hobbyist literature, commercial suppliers and organizations to support growing networks of enthusiasts.

Early in the century, novels for adolescents had combined modernist themes in a quick and exciting read. The original *Tom Swift* series (1910–41), conceived by American writer

[4] On the visual representation of modernity, see Wilson, R. G., D. H. Pilgrim and D. Tashjian, *The Machine Age in America: 1918–1941* (New York: Brooklyn Museum, 1986).

[5] On related themes, see Hughes, T. P., *American Genesis: A Century of Invention and Technological Enthusiasm* (New York: Viking, 1989).

and publisher Edward Stratemeyer (1862–1930) churned out mysteries that mixed invention, clear thinking, adventure, industrial secrecy and wondrous capabilities.[6] Several thousand titles influenced generations of American children and young adults.[7]

Magazines also captured technical enthusiasts. A seminal American publisher, Hugo Gernsback (1884–1967), chronicled invention, popular science and science fiction via some four dozen magazines for technical enthusiasts, several of which continue into the twenty-first century.[8] This publishing model was repeated on a smaller scale in other countries and helped produce a generation of active technical amateurs.[9]

Gernsback's publications extended the visual grammar of modernity (Section 3.6). Lurid covers depicted technologies of the future, confidently forecasting how television, air travel and spacecraft would be experienced. This aesthetic—combining visual shock with the seduction of imagined technology—shaped viewers' tastes and expectations.

Fiction and specialist magazines were supplemented by the promotion of popular science. The rising appeal of science and technology was sensed by a nascent organization, Science Service, in 1921. Its news syndication service and periodical, *Science News Bulletin*, were dedicated to the accurate reporting of science. An important secondary aim was soon added: to support the growth of new technical hobbies that provided hands-on experience with scientific culture.

One of the first campaigns of Science Service was its promotion of amateur radio. Science Service disseminated the information about building radios and forming clubs of radio enthusiasts via feeds to newspapers and popular periodicals such as *Good Housekeeping, Harper's Magazine* and *Popular Science Monthly.*[10] Its Director linked hands-on scientific insights to current events:

[6] Molson, F. J., 'The boy inventor in American series fiction: 1900–1930', *Journal of Popular Culture* 28 (1994): 31–48; Dizer, J., *Tom Swift & Company: 'Boys' Books' by Stratemeyer and Others* (Jefferson, NC: McFarland, 1982); Johnson, D., *Edward Stratemeyer and the Stratemeyer Syndicate* (New York: Twayne, 1993). A second series of novels (1954–71) featured the character's son, Tom Swift Jr, and updated inventions, resonated with children of the baby boom who took up holography.

[7] The successful format fitted the American cultural and political landscape well. E. F. Bleiler argues for the capitalist underpinnings of such juvenile fiction into the twentieth century, noting that the *Tom Swift* stories, like others of the genre, emphasized intelligence and hard graft—'economic parables', he suggests, about earning a living as much as about escapist adventure. This made the novels relevant to their readers' dreams of career success and influence (Bleiler, E. F., 'From the Newark Steam Man to Tom Swift', *Extrapolation* 30.2 (1989): 101–16; quotation p. 112. See also van der Osten, R., 'Four generations of Tom Swift: ideology in juvenile science fiction', *The Lion and the Unicorn* 28 (2004): 268–83). I argue more generally that technical amateurism during the twentieth century was closely linked to modernist ideals.

[8] Ashley, M., *The Gernsback Days* (Holicong, PA: Wildside Press, 2004); Massie, K. and S. D. Perry, 'Hugo Gernsback and radio magazines: an influential intersection in broadcast history', *Journal of Radio Studies* 9 (2002): 264–81.

[9] In Britain, a similarly prolific and influential enthusiast was Frederick J. Camm (1895–1959). Camm's publications promoted his enduring interests in technical hobbies such as aviation, radio, automotive engineering and, later, television. See Cullingham, G. G., *F. J. Camm, The Practical Man* (Windsor, UK: Thamesweb Publishing, 1996). Comparable publications and enthusiasts emerged elsewhere.

[10] Bureau of Standards, 'Construction and operation of a very simple radio receiving equipment', Letter Circular LC 43, 16 Mar 1922; US Dept. of Agriculture and State Agricultural Colleges, 'Cooperative extension work in agriculture and home economics—boys and girls radio clubs', 1922.

> Science Service in its early days pioneered in giving newspaper readers accurate and understandable instructions for building radio sets. When Lindbergh flew it told how to build model airplanes. When radiovision became experimental it described the construction of a radiovisor.[11]

Stronger support for scientific amateurism—and eventually for amateur holography itself—came from magazines aimed at more educated and aspirational audiences. *Scientific American* magazine defined and nurtured this cohort of enthusiasts, beginning with a regular column on amateur telescope-making. Founded in 1845, the periodical had been dedicated to new inventions but, from 1921, was reoriented as a magazine of popular science.

One of its editors, Albert G. ('Unk') Ingalls (1888–1958) described his promotion of amateur telescope making in the USA as 'sheer accident', although it appears to have been more a case of tactical publishing.[12] Browsing an account of a group of amateur Vermont telescope makers mentored by astronomer Russell Porter, Ingalls was intrigued enough to build his own telescope and to publish an article in *Scientific American* in 1925 about the group. Over 300 readers responded with requests for further information so, beginning the next year with Porter's assistance, Ingalls began a series of monthly articles, 'The Back Yard Astronomer' in the periodical. He and the magazine brought out a slim book, *Amateur Telescope Making*, the following year, and expanded it steadily to become a substantial set of tomes over the following three decades. Ingalls credited the publications with fomenting communities of 'scientifically minded persons' who showed 'keen enthusiasm, sometimes almost fanatical' for the growing hobby.[13]

Telescope-making attracted technical enthusiasts—'eager workers, young and old, skilled and less skilled, men and women (several of these)', but Ingalls argued that such amateur scientific activities could appeal to wider modern audiences:

> It exacts intelligence; requires patience and sometimes dogged persistence in order to whip the knotty but fascinating problems which arise; demands hard work—is not dead easy; and compels the exercise of a fair amount of handiness—enough to exclude the born bungler but no more than is possessed by the average man who can 'tinker' his car or the household plumbing, or dissect and wreck a watch.

The product of such a devoted worker would be 'a valuable scientific instrument which places him on the threshold of astronomy and astrophysics, perhaps the most romantic branch of modern science'.[14]

[11] Davis, W., 'Make your own telescope', promotion letter to newspaper editors, n.d., 1930, Smithsonian Institution Archives RU7091, Box 120, folder 9. Science Service records are archived at the Smithsonian Institution Archives, Washington DC, henceforth abbreviated SIA.

[12] Ingalls, A. G. to B. W. Powell, letter, 10 Apr 1953, ACNMAH 0175 Box 8, folder 2.

[13] Ingalls, A. G., *Amateur Telescope Making* (New York: Scientific American, 1933 (3rd edition)), quotations pp vii–viii, x; see also Ellison, W. F. A., *The Amateur's Telescope* (Belfast: Carswell & Son, 1920).

[14] Ingalls, A. G., *Amateur Telescope Making* (New York: Scientific American, 1933 (3rd edition)), quotations pp viii–ix.

By uniting isolated individuals across the continent and beyond, Ingalls' columns knitted together a virtual community of enthusiasts. Like contemporary amateur photographers and later amateur holographers, such individuals had an unusual mix of competences. They nurtured the artisanal skills of mirror grinding and telescope building alongside scientific competences such as celestial mechanics and patiently systematic astronomical practice.[15]

This nurturing of individual experimenters was paralleled by rising support from institutions. Science Service championed an offshoot, Science Clubs of America and an associated event, the Science Talent Search, as a means of nurturing young scientists for national benefit.[16] With sponsorship from Westinghouse Electric and Manufacturing Company, the clubs and prize event spread quickly. Founded the year that America entered the war, they linked amateur science to national purpose:

> The world has increasing need of reasoning and creative scientists. We must make the world a better place to live. Science is a prime agency in rebuilding civilization and outmoding war, just as it is today a powerful means of fighting to save our way of life and bring victory. Young men and women who are the members of the Science Clubs of America will help do this continuing task of carrying the torch of knowledge and investigation.[17]

This progressive vision of science was promoted by Science Service's *Science News Letter* and radio programmes (Figure 7.1).[18]

Such enthusiasms were further extended by the postwar attention to science and the new medium of television.[19] *Watch Mr Wizard*, a popular American programme between 1951 and 1965, portrayed a science hobbyist demonstrating his latest home experiments to visiting children. Within three years of its first airing, the show was telecast on some ninety stations. A growing network of 'Mr Wizard science clubs' attracted primary-school students across the USA and Canada.[20]

These initiatives for grass-roots science were subsumed within wider events. The wartime successes in government-funded research and development, combined with

[15] On Ingalls' influence, see Williams, T. R., 'Albert Ingalls and the ATM Movement', *Sky and Telescope*, 18 February 1991: 140–2. In Britain, a similar rise in amateur science was later supported by Sidgwick, J. B., *Amateur Astronomer's Handbook* (London: Faber & Faber, 1955).

[16] Marx, J., 'How the "Queen Science" lost her crown: a brief social history of science fairs and the marginalization of social science', *Sociation Today* 22 (2004): http://www.ncsociology.org/sociationtoday/v22/science.htm.

[17] Davis, W., '*Adventures in Science*: Fourth Annual Science Talent Search', CBS radio script, 17 Feb 1945, SIA RU7091, Box 391, folder 1.

[18] See Rhees, D. J., *A New Voice for Science: Science Service Under Edwin E. Slosson, 1921–1929*, MA dissertation thesis, History, University of North Carolina at Chapel Hill (1979); Terzian, S. G. and L. Shapiro, 'Corporate Science Education: Westinghouse and the Value of Science in Mid-Twentieth Century America', *Public Understanding of Science* (2013): 1–20.

[19] Lafollette, M. C., *Science on American Television: A History* (Chicago: University of Chicago Press, 2013).

[20] See also Herbert, D., *Mr Wizard's Science Secrets* (USA: Popular Mechanics Co., 1952). Rather like a junior version of 'The Amateur Scientist', the book and programme sought to encourage curiosity and practical skills while learning science: '. . . the part of science that's fun to investigate for yourself right at home. Milk bottles are your flasks, glasses are your beakers and the whole house is your laboratory' (p. 5).

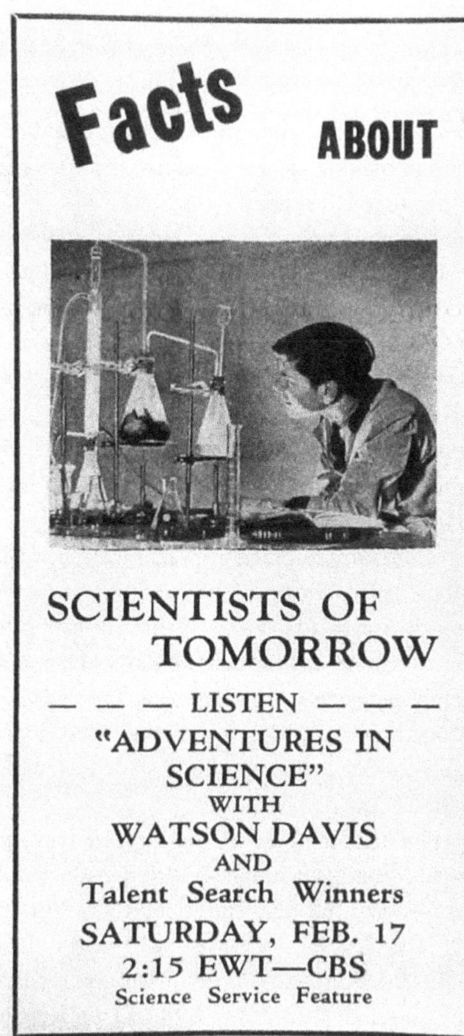

Fig. 7.1 Promoting amateur science (Science Service, Radio programme advertisement, 'Facts about scientists of tomorrow' (Science Service: Washington DC, 1945), 1, from SIA RU7091—Science Service records, Box 391, folder 1).

postwar concerns about international markets and Cold War competition, encouraged postwar governments to promote science education at the national level.[21] In August 1949, the first Soviet atomic weapon test received blanket coverage in American newspapers and popular magazines, triggering rhetoric that escalated through the decade:

[21] Wang, J., *American Science in an Age of Anxiety: Scientists, Anticommunism, and the Cold War* (Chapel Hill: University of North Carolina Press, 1999).

> If high schools would take as much pride in outfitting chemistry and physics labs as they do in outfitting their football teams . . . The United States would be well on its way to solving the serious threat to its survival posed by the alarming shortage of engineering and scientific man-power and the growing threat of Soviet technological superiority.[22]

Caught up in the rising cultural tide towards popular science and technology, and hastened by anxieties about Soviet competition in making atomic weapons, the vogue for science clubs and science fair competitions expanded through the 1950s.[23]

Technical enthusiasms were encouraged further by a growing industry supplying affordable components for technical hobbies. The firms co-evolved with their customers. Stocks shaped hobbyist projects, and project goals encouraged new sources of supplies. The most successful of these firms satisfied disparate and rapidly-evolving hobbies.

In the New York City district later occupied by the World Trade Center, 'Radio Row' had catered for enthusiasts of radio, television and electronics. Such surplus suppliers became a feature of many cities with the introduction of radio broadcasts internationally from the early 1920s. The largest among them printed catalogues for a growing mail-order clientele, and became the model for later electronics supply businesses.

A variant business model was the Heath Company, founded in 1912. Heath had specialized in aircraft kits and accessories until the end of the Second World War, attracting a variety of technical amateurs by selling products that adapted to any level of expertise: either a complete aircraft, the airframe minus engine, a kit of aircraft parts, the engine alone, an engine modification kit, or merely blueprints. After the war, the company broadened its market by selling electronic kits of war surplus components. These collections and carefully detailed instructions could up-skill postwar enthusiasts of electronics and also yield modern consumer technologies at low cost.

Specialist companies trading in war surplus technologies were a treasure-trove for more adventurous technical hobbyists. These component suppliers served distinct audiences. Some postwar hobbyists became skilled and independent enough to adapt and design their own projects.

The most important of these for hologram enthusiasts was Edmund Salvage Corporation, founded in New Jersey by Norman W. Edmund (1916–2012). Edmund, an amateur photographer, resold wartime production of 'seconds' to amateurs seeking inexpensive optical components.[24] Surplus lenses, mechanics and electrical supplies appealed to hobbyist makers of telescopes, microscopes and other optical devices. The company later packaged low-cost collections of components and demonstration supplies for schools,

[22] The Science Talent Institute, 'Teen-age scientists offer cures for threats to U.S. technological survival', press release, 5 Mar 1956, SIA RU7091, Box 330 folder 5.

[23] Terzian, S. G., Science Education and Citizenship: Fairs, Clubs, and Talent Searches for American Youth, 1918–1958 (New York: Palgrave Macmillan, 2013).

[24] Other postwar suppliers of optics parts, often advertising in *Scientific American* alongside the 'Telescoptics' and later 'Amateur astronomer' sections, included the Associated Surplus Company, F. W. Ballantyne, Columbo Trading Company, A. Cottone, A. Jaegers and Harry Ross, each based in either California or New York.

and was renamed the Edmund Scientific Company in 1951. From the 1960s, the mail-order company expanded into meteorology, chemistry and science gadgets to capture the rising tide in amateur science and technology.

Such companies proved successful and long-lived for at least three reasons. First, they exploited, and later developed, sources for low-cost stock. By tapping war-surplus supplies they channelled specialized components to creative amateurs. Once captured, those amateurs became a reliable market for further manufacturers that could serve the supply companies with amateur-oriented products. Second, their self-published pamphlets and books coached enthusiasts in the requisite technical skills. And third, the companies offered routes of intermediate difficulty through the technologies they promoted: hobbyists could buy raw components, step-by-step kits or ready-made products by mail-order, linking them to a dispersed and graded community of like-minded enthusiasts.

Publishers adapted to the new emphases of postwar amateurs. In 1948, the struggling *Scientific American* magazine was relaunched by new owners and pitched at a more refined audience, aiming to stretch its readers: 'The function of *Scientific American* is to present the current and historic progress of science to an audience of educated laymen'.[25]

Combined with the wider cultural enthusiasm for science heightened by the war and *Scientific American* itself, Ingalls's column was retitled 'The Amateur Scientist' in 1952 and, until his retirement in 1955, was largely ghost-written by one of the founding editors of the revamped magazine, Clair L. ('Red') Stong (1902–76). Stong's breezily-drafted account for a would-be contributor to the column revealed postwar attitudes about popular science in American publishing:

> Got into *Scientific American* this way: Gerry Piel, scion of a local beer maker; Dennis Flanagan, former science editor of *LIFE*; Leon Svirsky, former science editor of *TIME* and I used to sit around various saloons, mumbling amiably into our beer the while.
>
> Science seemed to make more sense to us than God. (Proof, doubtless, that we are much in need of the analyst's couch.) Seemed to us that the Common Man could come more effectively to grips with his social problems if he knew a bit more about the cultural force which (in our opinion) above all others currently shapes them—science.[26]

An electrical engineer for Westinghouse, Stong was initially unsure whether scientific hobbyists existed in large numbers, and just how they would relate to professional workers:

> From occasional references in professional journals and related sources, we have gathered the impression that much worthwhile work is being conducted by gifted laymen who have turned to science as an avocation ... We believe the researches of these workers merit regular publication in our pages. Not only would publication aid the amateur in gaining professional recognition but, more importantly, it would encourage a broader public understanding and participation in science. Accordingly, we are at present considering the establishment of a department of amateur science as a regular feature of the magazine.[27]

[25] *Scientific American* magazine, 'An author's guide to *Scientific American*,' New York, 1949.

[26] Stong, C. L. to H. Morgenroth, letter, 2 Feb 1955, ACNMAH 0012 Box 4, folder 1.

[27] Stong, C. L. to H. H. Larkin, letter, 11 Sep 1951, ACNMAH 0012 Box 1, folder 1.

Soliciting the first year's articles from personal contacts and recommendations of Science Talent Search winners, the avuncular Stong soon found readers contacting the magazine eager to describe their own scientific achievements. Bolstered by the postwar availability of war surplus electronics, optics and mechanical parts, his articles—continuing monthly for the next twenty-two years—described the apparatus and investigations undertaken by a generation of amateurs, ranging from an electronic seismograph to a home-made atom-smasher.[28] And, like Science Service, the Amateur Scientist Department was attuned to its times. The launch of Sputnik in October 1957 led to an article just three months later on how to view and time it.[29]

In introducing a book collection of the articles, Vannevar Bush (1890–1974)—electrical engineer, university administrator and wartime overseer of the Manhattan Project—emphasized the link between amateur enthusiasms, science and society:

> There are lots of amateur scientists, probably a million of them in this country. The Weather Bureau depends on some 3000 well-organized amateur meteorologists. Other groups observe bird and insect migrations and populations, the behavior of variable stars, the onset of solar flares, the fiery end of satellites, earth tremors, soil erosion, meteor counts, and so on . . . there are amateurs who are truly masters of their subjects, who need take a back seat at no professional gathering in their field. It was an amateur who discovered Pluto, and an amateur who was primarily responsible for the development of vitamin B1.

While Bush's commendation carried a hint of faint praise, Stong himself addressed his readers as more autonomous individuals:

> I have supposed that you revel in your simian heritage of curiosity. You take boundless delight in finding out what makes things tick, whether the object of your interest has been fashioned by nature or man. Second, you are an inveterate tinkerer. You love to take organized structures apart and put them together again in new and interesting ways—be they rocks, protozoa, alarm clocks or ideas. Third, you can usually drive a nail home on the first try, put a fairly good edge on a knife, and manipulate a Bunsen burner without broiling your thumb. In short, I assume you are an Amateur Scientist.[30]

[28] Typical columns included Stong, C. L., 'The Amateur Scientist: how to compute the orbits of space vehicles', *Scientific American*, 220 (1), Jan 1969: 123–30 and Johnston, S. F. and C. L. Stong, 'The Amateur Scientist: a high school student builds a recording spectrophotometer', *Scientific American*, 231 (1), Jan 1975: 118–25.

[29] New York Times News Service, 'Satellite search a hobby for home: Los Alamos scientists build watching post based on simple short wave unit', *New York Times*, 20 Oct 1957, 8; Stong, C. L., 'The Amateur Scientist: mostly about how to study artificial satellites without complex equipment', *Scientific American*, 201 (1), Jan 1958: 98–109.

[30] Stong, C. L. (ed.), *The Scientific American Book of Projects for the Amateur Scientist* (New York: Simon & Schuster, 1960); Introduction by Vannevar Bush; Preface by C. L. Stong; quotations pp xviii–xix and p. xxi, respectively. Bush's vision of amateur activity has more recently been dubbed 'citizen science' (Edwards, R., 'The "citizens" in citizen science projects: educational and conceptual issues', *International Journal of Science Education, Part B* 4 (2014): 376–91). Unlike the technical enthusiasms explored in this book, citizen science is typically hierarchical, with groups of amateurs subordinate to professional scientists.

As represented by Bush and Stong, the traits of the amateur enthusiast were diverse but intense. Within the pages of *Scientific American* magazine amateurs could co-exist with professional scientists without hierarchy or condescension.

Hologram enthusiasts gained their first foothold on this framework of technical amateurism established in previous decades. Just as photography, radio-building and telescope-making had generated enthusiasms and skills, and publishers, supply companies and mentoring schemes had supported them, new scientific pastimes were actively nurtured during the 1950s and sixties. Postwar culture encouraged new technical enthusiasms that were tightly aligned to the *zeitgeist* of modernity and progress, a cultural embedding at precisely the right time to encourage a generation of hologram-makers, too.

7.3 HOLOGRAMS FOR AMATEURS

> *Holograms are the 3-D TV of Tomorrow.*
>
> Edmund Scientific catalogue, 1975

Prior cultural underpinnings, hobby networks and commercial suppliers provided a ready-made template for hologram enthusiasts. Amateurs, in turn, proved to be an important cultural channel for transmitting holograms to a wider public. They scaled-down 'big science' to make it feasible for technically-adept individuals. Just as importantly, they translated covert and unattainable science into a more recognizable product. Lasers and holograms could escape closed laboratories and find a home in basement workshops.

The uptake of holograms mirrored the amateur enthusiasms shown for earlier technologies. A popular literature encouraged curiosity and engagement; technical hobbyists extended and refined their skills and interests; firms identified new demands for their products, and developed new ones. The result was a growing band of enthusiasts who shaped the subject and subsequent audiences.

As boys' novels and hobbyist magazines had inspired the previous generation, new publications and products attracted and motivated hologram hobbyists. Information about holograms spread during the mid-1960s, allowing alert enthusiasts to pick up intriguing details, obtain samples and even to experiment with making them.

Popularizations, products and how-to articles about holograms reached wider publics from 1965. That autumn, some 18 months after the first announcements and a string of presentations at American engineering conferences, Leith and Upatnieks wrote an article for *Scientific American*. Pitched, in the accessible style of the magazine, at professional and para-professional audiences—scientists, engineers, technicians and science enthusiasts—the article triggered a wave of active interest.[31]

[31] Leith, E. N. and J. Upatnieks, 'Photography by laser', *Scientific American*, 212 (6), Jun 1965: 24–36.

The first article about holograms for hobbyists also appeared that year, dedicated not merely to viewing the rare objects, but to creating them at home.[32] It appeared in the 'Amateur Scientist' Department of *Scientific American*, which by then had a reputation for promoting autonomous amateur science of a high standard.

The editor of the Department, Red Stong, noted that he had been 'all set to start experimenting with holograms' when a correspondent—a furniture company owner on Staten Island—wrote to the magazine in May 1966 to propose the article. Stong's reply balanced the need for variety with the demands of ardent amateurs:

> 'Hobbiests [sic] in each field are entitled to an occasional item for the 75 cents. Accordingly, I try to space articles in a given field about 12 months apart. E.g. I told em how to build a primitive HeNe laser in September 1964 and how to tinker with it in December 1965. I should suppose that we might tell them how to improve it toward the end of this year, and how to make holograms toward the end of 1967. In a sense, this is too bad because I have scores and scores of letters asking how to make holograms.[33]

His long-time illustrator, artist Roger Hayward, agreed the subject was timely but 'at the present I tend to class holography with fiber optics, very interesting but no one has found much important use for it yet'.[34] The article, appearing in early 1967, appealed to hard-core amateurs adept at building laser tubes and power supplies, and willing to stretch to optical tinkering (Figure 7.2).

The account notably omitted a photograph of the contributor's resulting holograms, which were judged by senior editors to be too poor in quality (Figure 7.3). Instead, it substituted pictures of one of the first holograms sold by Edmund Scientific Corporation.[35]

The next article on holograms was also an atypical cheat: Stong had been door-stepped by a publicist at IBM to publish an article by one of its engineers that had been written for university student labs. Stong rewrote the account in the house style, but it lacked the innovative amateur spirit of the typical column. Both articles failed to generate many responses from readers of *Scientific American*: the first was too demanding—requiring not just a home-made laser, but also optics expertise—and the second suited a well-equipped university teaching lab but not amateurs obtaining their parts from war-surplus component catalogues.[36] Both hologram articles misrepresented amateur achievements while raising their aspirations.

Over the following decade, the magazine periodically published more accurate columns detailing laser construction and hologram-making based on correspondence from

[32] Heumann, S. M. and C. L. Stong, 'The Amateur Scientist: how to make holograms and experiment with them or with ready-made holograms', *Scientific American*, 216 (2), Feb 1967: 122–8.

[33] Stong, C. L. to S. Heumann, letter, 12 Jul 1966, ACNMAH 0012 Box 23, folder 7, henceforth SIA.

[34] Hayward, R. to C. L. Stong, letter, 10 Dec 1966, ACNMAH 0012 Box 23, folder 7.

[35] Corey, L. E. to C. L. Stong, letter, 14 Nov 1966, ACNMAH 0012 Box 23, folder 7.

[36] Rowe, D. F. to C. L. Stong, letter, 4 Feb 1971, ACNMAH 0012 Box 30, folder 6; Dickson, L. D. and C. L. Stong, 'The Amateur Scientist: stability of the apparatus: insuring a good hologram by controlling vibration and exposure', *Scientific American*, 224 (7), Jul 1971: 110–7.

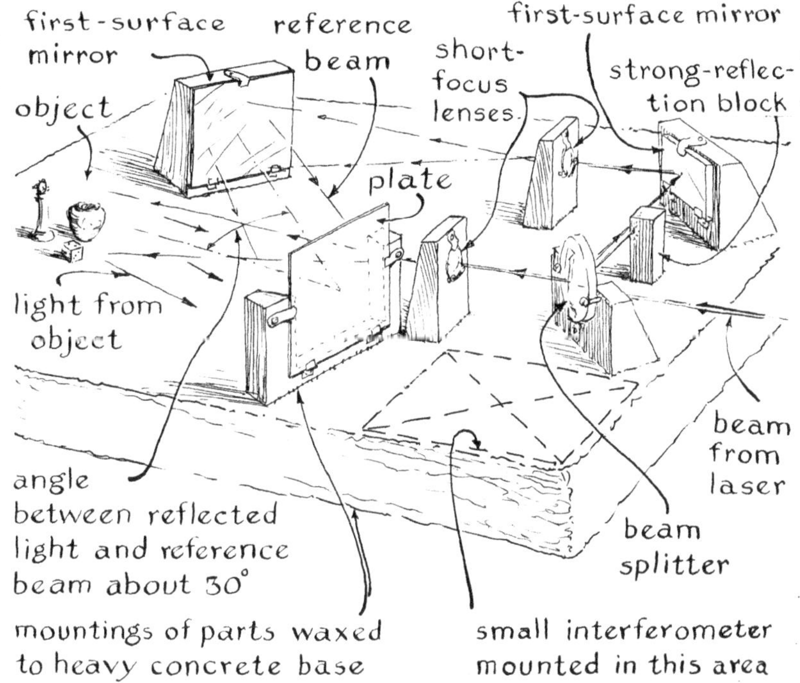

Fig. 7.2 Making your own holograms in Amateur Scientist (original illustration by Roger Hayward, 1966, ACNMAH 0012, Box 23, folder 7, by kind permission of Miriam and Jim Kramer, Roger Hayward estate).

home tinkerers.[37] *Popular Science* described the methods too, sandwiched between articles on building a basketball backboard and 'what you should know about torque wrenches'. Holograms, it seemed, had been house-trained.[38]

Holograms began to circulate, too. A technician at another of the Willow Run laboratories, Don Gillespie (b. 1934), began making holograms in his basement at home.

[37] Stong, C. L., 'The Amateur Scientist: how a persevering amateur can build a gas laser in the home', *Scientific American*, 211 (9), Sep 1964: 127–34; Stong, C. L., 'The Amateur Scientist: more about the home-made laser', *Scientific American*, 213 (12), Dec 1965: 106–13; Stong, C. L., 'The Amateur Scientist: how to construct an argon gas laser with outputs at several wavelengths', *Scientific American*, 220 (2), Feb 1969: 118–25; Stong, C. L., 'The Amateur Scientist: how to construct a dye laser', *Scientific American*, 222 (2), Feb 1970: 116–23; Levatter, J. and C. L. Stong, 'The Amateur Scientist: a carbon dioxide laser is constructed by a high school student in California', *Scientific American*, 225 (3), Sep 1971: 218–24; Iizuka, K. and C. L. Stong, 'The Amateur Scientist: holograms with sound and radio waves', *Scientific American*, 226 (11), Nov 1972: 120; Stong, C. L., 'The Amateur Scientist: how to construct an infrared diode laser', *Scientific American*, 227 (2), Mar 1973: 114–21; Stong, C. L., 'The Amateur Scientist: how to construct a nitrogen laser', *Scientific American*, 229 (6), Jun 1974: 122–7; Walker, J., 'The Amateur Scientist: how to stop worrying about vibrations and make holograms viewable in white light', *Scientific American*, 260 (5), May 1989: 134–7.

[38] Benrey, R. M., 'Make holograms at home – at a budget price', *Popular Science*, 198 (1), Jan 1971: 84–6.

(a)

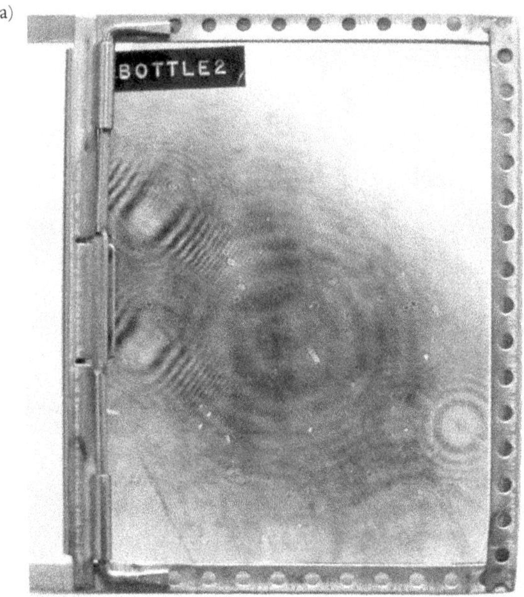

(b)

Fig. 7.3 Images of (a) one of the first amateur holograms and (b) its reconstructed image (Sylvain M. Heumann, 1966, with kind permission of W. Heumann).

Through what were probably the first advertisements for holograms in early 1966, he recalls selling a half-dozen for $100 each.[39] Another Willow Run employee, Lloyd Cross (1934–2015), spawned the first laser company and began experimenting with holograms about a year later. The same year, yet another of their peers, Frank Denton (b. 1935), began manufacturing holograms covertly at the University of Michigan to sell via his start-up company, Photo-Technical Research (PTR). Using its own facilities, the company produced the first large order for holograms that year: some fourteen thousand Leith–Upatnieks holograms to be included as inserts in the July 1967 trade magazine *Laser Focus*. Targeted at American engineers, the holograms appear to have disappeared without trace or wider impact.[40]

The most widely available, however, were the half-million holograms manufactured by the Conductron Corporation of Ann Arbor. The holograms, published with the *World Book Science Annual* in 1967, were accompanied by a red plastic filter and instructions for building a suitable illuminator from the filter and a slide projector (Figure 7.4). While still reaching an uncertain and largely unresponsive audience, such

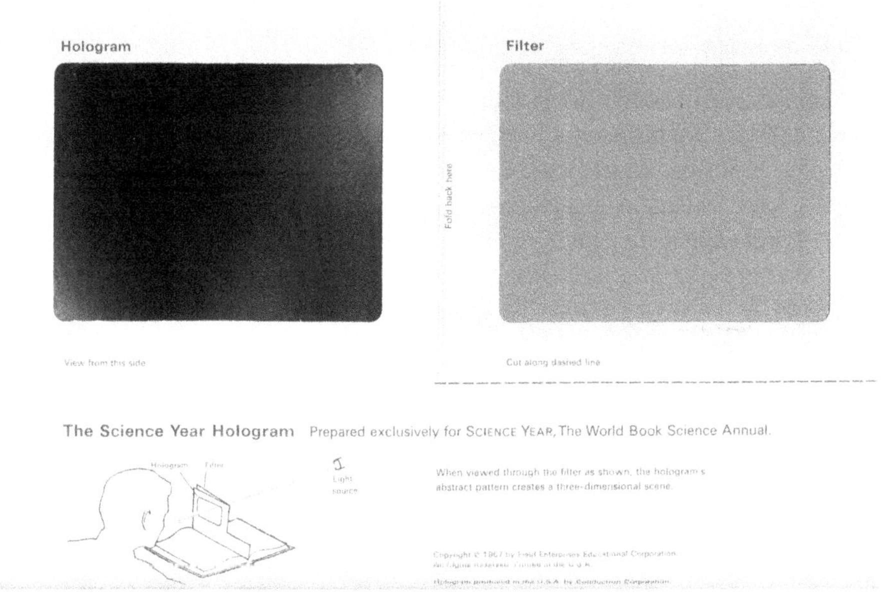

Fig. 7.4 First widely distributed hologram (left) with red viewing filter (right) (Cochran, Gary D. and Robert D. Buzzard, 'The new art of holography', in: *Science Year: The World Book Science Annual 1967* (Chicago: Field Enterprises Educational Corp, 1967), pp 200–11; hologram insert is pp 209–10).

[39] Gillespie, D. and JODON Ltd to C. L. Stong, letter, 11 Nov 1966, ACNMAH 0012 Box 23, folder 7; Gillespie, D. to SFJ, interview, 29 Aug and 4–6 Sep, 2003, Ann Arbor, MI, SFJ collection.

[40] Cross, L. to SFJ, email, 25 Oct 2003, SFJ collection; Denton, F. to SFJ, telephone, 1 May 2003, SFJ collection; 'The first hologram ever distributed to a mass audience', *Journal of the SMPTE* 76 (1967): 707.

Ann Arbor products provided the first opportunities for most Americans to see holograms into the 1970s.[41]

To promote a commercial market for holograms, Conductron also distributed free samples to advertisers and produced demonstration runs for promising products. Some 35,000 holograms, for example, were manufactured for the Hoffman La Roche Company in 1969 to advertise their pharmaceuticals for colon treatments. Although few of the doctors who received them were influenced in their purchasing habits, many of the holograms percolated down to their more enthusiastic friends and relatives. A similar attempt to reach wider audiences was the impressive reflection (Denisyuk) hologram bound into the book of the 1971 Nobel Prize speeches (which included the Physics Prize for Denis Gabor's invention of holography). But, given the exclusivity of the publication, only the most motivated members of the public were likely to view it. The venue of viewing—likely in a public library under fluorescent lighting—would have further limited the visual impact. Personal access to holograms was thus a rare privilege during the 1960s, and creating them demanded dedication.[42]

As suggested by the entrepreneurial activities of Willow Run technologists, amateur holography during the 1960s relied on unusual technical competences. The hobbyists targeted by Stong's column were counselled that prior experience with electronics and metalwork would be needed to build and operate a gas laser—the only option available for middle-class enthusiasts. Some aptitude in optics was needed to successfully mount and adjust lenses and mirrors for holography. Mechanical adeptness could solve the ubiquitous problems of vibration and the consequent failure of exposures when recording holograms. Familiarity in a photographic darkroom was at least equally important. Chemically developing holograms was much like processing regular photographic film, but there were paradoxical aspects beyond the ken of amateur photographers. An excellent hologram could be achieved, for instance, even with a scarcely exposed or nearly black photographic plate.

The properties important to making a successful hologram were entirely invisible. Infinitesimal movement would destroy the recording, even when appearing properly exposed and developed. The type of photographic film or plate that was used could dramatically affect the set-up; most consumer films and optical arrangements didn't work at all. The coherence of the laser beam—a subtle characteristic relating to how well the emerging light-waves were in step—was equally crucial. Mechanical strain or

[41] Gillespie, D. to SFJ, interview, 29 Aug and 4–6 Sep, 2003, Ann Arbor, MI, SFJ collection; Cochran, G. D. to SFJ, interview, 6 and 8 Sep 2003, Ann Arbor, MI, SFJ collection; Cochran, G. D. to SFJ, interview, 6 and 8 Sep 2003, Ann Arbor, MI, SFJ collection; Cochran, G. D. and R. D. Buzzard, 'The new art of holography', in: *Science Year: The World Book Science Annual 1967* (Chicago: Field Enterprises Educational Corp, 1967), pp 200–11. Gillespie formed Jodon Engineering to supply lasers and custom optics for the new market, and Cross was a principal in Trion Lasers, the first laser company. Conductron Corporation, another spin-off of technical staff of the Willow Run Labs, was responsible for most of the holograms manufactured during the 1960s and early 1970s.

[42] Charnetski, C. to SFJ, interview, 3 Sep 2003, Ann Arbor, MI, SFJ collection; *Les Prix Nobel En 1971* (Stockholm: Imprimerie Royale P. A. Norstedt & Söner, 1971); hologram is between pp 176–7.

temperature shifts of the subject and mirrors could fatally smear the recording while leaving its beam seemingly unchanged. The range of light and dark (dynamic range) and contrast of the hologram image could depend on the balance between different beams of laser light lighting the apparatus. And the pragmatic understandings of imaging optics that were familiar from wartime experiences with binoculars, cameras, signalling lamps and transits were largely irrelevant. Making holograms required no focusing, no corrected lenses and no alignment; instead, crude optical components with clean surfaces seemed to do the trick. Such arcane properties—equally baffling to many of the professionals of the period—made the creation of holograms frustrating and expensive.[43]

Enthusiasts with particular backgrounds had a head-start. Few had all the requisite skills. More commonly, a familiarity with electronics and mechanical workshop skills provided essential grounding, with darkroom and optical work acquired along the way. As Leith and Upatnieks had found at Willow Run Labs, disciplinary boundaries had obscured and delimited the new field of holography. Technical pastimes were free of institutional secrecy but often found themselves isolated by their intense enthusiasms and physical separation. The required interdisciplinary skill set could be developed more quickly by sharing experiences with other hobbyists, but during the 1960s such avenues were rare. 'The Amateur Scientist' provided a link between separate fields, and amateur photography, electronics and astronomy enthusiasts gradually extended their pastimes to new territories. For amateur radio hobbyists, building and even designing power supplies, filters and oscillators had become an expanding skill as new postwar electronic devices were conceived. Amateur telescope-makers were versed in mirror-fabrication, mounts and sensitive but robust mechanical adjustments in ways that their professional counterparts often were not. These merging amateur cultures assumed a recognizable form when they turned outward. From the 1970s, the new practical art of hologram creation could be taught by instructors who had started from earlier hobbies.

As with prior waves of photography and electronics hobbyists, supply companies recognized a potential market in hologram enthusiasts. The Willow Run technologists focused on products for their peers, but Edmund Scientific was particularly well aligned with the interests of technical hobbyists and schools (see Section 8.4). For their clientele inspired by popular science, electronics hobbyist magazines and amateur telescope-making, the company could now add new and more exciting products: light sources to view and even make holograms, and holograms themselves in a growing variety.

The first holograms sold by Edmund were transmission holograms in 1966, changing to a higher quality hologram of a chess scene manufactured by Frank Denton's PTR the following year.[44] The hologram products of other companies were added through the 1970s and 80s and promoted to whet the appetites of hobbyists and early adopters. The 1977 introduction of animated holograms from San Francisco's Multiplex Company, for

[43] Cochran, G. D. to SFJ, interview, 6 and 8 Sep 2003, Ann Arbor, MI, SFJ collection; Leith, E. N. to SFJ, interviews, 29 Aug—12 Sep 2003, Ann Arbor, MI, SFJ collection.

[44] Corey, L. E. to C. L. Stong, letter, 14 Nov 1966, ACNMAH 0012 Box 23, folder 7.

example, were described as 'Moving Images of the Future [. . .] a forerunner of future holographic motion pictures' (see Chapter 8).[45]

The company fostered new manufacturing enterprises, too. Metrologic Instruments, located near Edmund Scientific in New Jersey, was founded in 1969 to supply laser kits to educators, initially through Edmund and later by retail to high school and college teachers as sales grew.[46]

The bar to amateurs, though, was still set high. The holograms invented by Leith and Upatnieks required a laser not just for creation but also for viewing. The red filter supplied in the *Science Year* book and first Edmund products resulted in a very fuzzy—albeit identifiably three-dimensional—image of chess pieces (Figure 7.5). Edmund Scientific had a better solution: war-surplus interference filters that transmitted a much narrower band of wavelengths from bright white sources, and so yielded sharper hologram images. For the more adventurous viewer, the company offered a low-pressure mercury lamp, a specialist product that had been familiar only to optical technicians and undergraduate physics labs in prior decades. These, too, provided acceptably sharp holograms. But, as with the interference filter approach, the illumination was dim and hologram viewing was confined to a darkened room.

The best solution, though, was a helium–neon (HeNe) laser. Invented in 1960, the HeNe was initially an expensive and finicky technology. Amounting to a variant of a neon lamp with delicate mirrors aligned at each end, the device could yield a continuous bright red narrow beam of light. By late 1962 relatively affordable and reliable commercial models were becoming available for industrial labs like Willow Run. Only after exploratory research provided a new market, though, did dropping prices allow schools and individuals to buy lasers.[47]

For most would-be hologram creators, ready-made lasers were the answer. Edmund introduced its first laser in 1970. It was relatively dim (0.5 mW), but adequate to record a hologram if exposed for tens of seconds. At a price of $498 (equivalent to some $3000 in 2015 currency) this essential component was out of reach for all but the most ardent enthusiasts. Alternatively, the company provided a 'laser optics kit' (laser tube, polished window and high-quality end-mirror) for $198, but this still needed a precision optical mount, power supply ($99.50) and housing.

The requirements for making holograms were even higher. The laser had to be stable over the duration of the exposure; if not, the interference patterns reaching the film would smear and ruin the recording. This required laser tubes to be mechanically and thermally stable. To the dismay of hobbyists, this could mean their bright new laser beam

[45] Edmund Scientific Catalog, 1978.

[46] As noted by Frank DeFreitas, there was a cluster of companies selling holograms, lasers and associated items in southern New Jersey and southeast Pennsylvania. Edmund and Metrologic were joined by the Holex Corporation (1972–82), a supplier of holograms to Edmund, Holographics Corporation (founded 1976), a maker of laser transmission (Leith–Upatnieks) holograms, and the Keystone Scientific Company (founded 1985), a supplier of holography kits.

[47] Bromberg, J. L., *The Laser in America, 1950–1970* (Cambridge, MA: MIT Press, 1991).

DOUBLE IMAGE HOLOGRAMS NOS. 41,091-41,094

These are transmission types of holograms and are viewed the same way as the transmission types, however, there are two separate objects on the hologram. To reconstruct these images, hold the film in the beam of the light source and turn it until the group of geometrical shapes appear in the air between the light source and the hologram. After you have these as clear as possible and with maximum brightness, rotate the film 180 degrees without moving it from the light beam. The image of the dinosaur will then appear. See Figure 4.

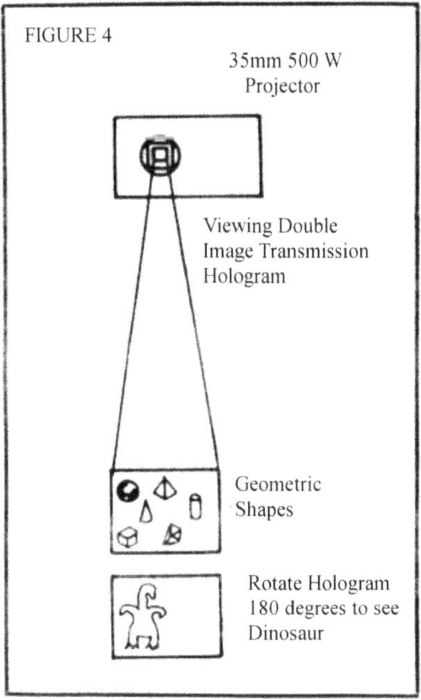

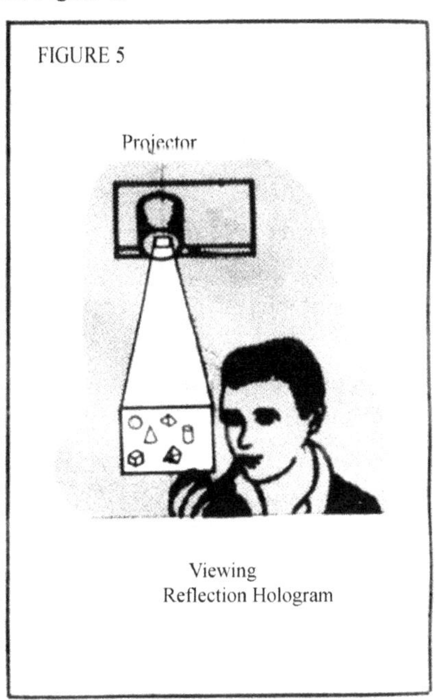

Fig. 7.5 How to view a hologram, 1960s style (Edmund Scientific Co. 'Information and instructions – holograms', p. 3, c1969. Re-lettered by J. Freeman, courtesy of Edmund Optics Inc).

had to be choked by a spatial filter, which allowed only one mode of the laser's oscillations to light the hologram set-up. Constructing a successful laser was neither trivial nor cheap.

War-surplus optical products were of relatively low quality but entirely adequate for generating high-quality holograms. 'Simple' (uncoated, single-element) lenses sufficed for laser light; clean but cheap microscope objectives served to expand the beam to illuminate or view and war-surplus mechanical mounts could be drafted in to steer the laser beam in the desired direction. Even so, individuals making holograms during the 1960s were likely to be technologists such as Lloyd Cross, Don Gillespie and Frank Denton who had access to institutional resources—or the acquaintances they mentored.

But once supply companies began to offer more affordable components and catalogues pitched at technical enthusiasts, amateurs were converted into experts who could vie with professionals. One of the most successful of them was Steve McGrew (b. 1945), who transformed a personal interest into a career as a hologram entrepreneur and inventor:

> I loved the accessibility of holography: if you could afford a few hundred dollars for a HeNe laser and some holographic plates, and if you were willing to learn to use a hacksaw and a drill and build components out of PVC pipes, scrap metal and a few precious lenses and mirrors purchased from the Edmund Scientific catalog, then you could become a holographer. I loved the fact that even non-technical people were attracted to holography and, in the process of building their own equipment, became expert in things that university physics professors perhaps should have known (but didn't).[48]

This sense of mastery of an arcane practice echoes the perspectives of mediaeval alchemists. Making holograms during the 1960s required solitary dedication to a black art, in which unwritten knowledge, weak lasers, insensible vibrations and novel chemical formulas combined to baffle neophytes. But secrets discovered in the dark rooms could trump professional investigators. Over the following decade, this search for the philosopher's stone turned into a quest shared by a growing band of amateurs. In the process, holograms were transmuted from a scientific wonder into an even more intriguing cultural product.

[48] McGrew, S., 'My reflections on holography', unpublished manuscript, 5 Dec 2004, SFJ collection.

8

Hologram Communities

So-called scientists who dabble
try their hand at science writing
...
popular science for the masses
crumbs of babble for the rabble
...
time's up
You know you got to
Run the hoodoo out of town.[1]

George Dowbenko, Taos, New Mexico, 1978

By the late 1960s, some five years after the first publicity about laser holograms, holography was still dominated by government and corporate funding. Exploratory American research—funded by the Department of Defense, NASA, National Science Foundation (NSF) and a cluster of corporations—inspired a handful of enthusiasts and hobbyists, but few could marshal the necessary resources or sustain independent activity. Holograms had aroused a growing popular curiosity about holograms and sparked a first wave of civilian enthusiasts, but also triggered a cultural backlash.

For certain audiences, holograms began instead to focus criticism of contemporary science. And, just as remarkably, holograms were taken up by members of such communities as a new model of science. The adoption of holograms by new communities increasingly explored cultural niches that were at odds with their origins. This chapter argues that these currents liberated holograms to play a more engaging role in popular culture over the following two decades.

The growing distance between 'big-science' holography and wider culture is illustrated by the swelling student protests during the 1960s. Dubbed 'the counterculture' by cultural historian Theodore Roszak, the movement challenged the established hierarchies of authority. Ann Arbor, Michigan, was point of origin not only of the Willow Run Laboratories (WRL) and modern holography, but also of the Students for a Democratic Society (SDS), founded in 1959 from the youth branch of the League for Industrial Democracy. SDS published a political manifesto, the 'Port Huron Statement' three years

[1] Dowbenko, G., *Homegrown Holography* (Garden City, NY: Amphoto, 1978), p. 9.

later that called for a more participatory form of democracy to address the wider social influences of racism, poverty, materialism, and militarism. It identified modern science as a product of, and contributor to, unjust political structures. In October 1963—as Leith and Upatnieks were beginning their first successful experiments with three-dimensional holography—a large student rally on campus protested the American military intervention in Indo-China. SDS membership at American universities mushroomed in 1966, when automatic student deferments for the draft were abolished.[2]

The SDS opposition to militarism, particularly through its Radical Caucus, focused increasing attention on universities engaged in classified research. In 1967, a sit-in at the University of Michigan Administration building specifically protested the university's classified research at facilities such as Emmett Leith's holography laboratory, which had been moved the year before from Willow Run to the town's North Campus. The protest helped to spearhead campaigns by students at over a hundred other American campuses over the following year against the war and militarily-funded research. During 1968, such dissent joined similar youth protests around the western world that more generally challenged established values, authoritarian government and the practice of modern science itself. One night that autumn, an uncredited detonation of dynamite destroyed the door and windows of Leith's Radar and Optics Lab (Figure 8.1).

Fig. 8.1 Bomb-damaged door and windows of Emmett Leith's holography lab, October 1968 (photo courtesy of C. Leonard).

[2] Roszak, T., *The Making of a Counter-Culture: Reflections on Technocratic Society and Its Youthful Opposition* (London: Faber and Faber, 1970); Unger, Irwin and Debi Unger, *The Movement: a History of the American New Left, 1959–1972* (New York: Dodd Mead, 1974). On the youth movement and the role of students at Ann Arbor, see also Breines, Wini, *Community and Organization in the New Left, 1962–1968: The Great Refusal* (New York: Praeger; J.F. Bergin, 1982) and Bunzel, John H., *New Force on the Left: Tom Hayden and the Campaign Against Corporate America* (Stanford, CA: Hoover Institution Press, Stanford University, 1983).

Campus unrest led the US Department of Defense to re-evaluate its funding of university research, and the growing controversy and declining income convinced University administrators to move the contract research off-campus to a lower profile and more vaguely-named organization, the Environmental Research Institute of Michigan (ERIM).[3] Ann Arbor's uncomfortable juxtaposition of holography researchers, on the one hand, and students opposing military-related research, on the other, was mirrored in two other locales: the Bay Area of California, home to Stanford University, and Cambridge on the opposite coast, where the Massachusetts Institute of Technology (MIT) had led wartime military research and spawned military contract companies along the 'Boston corridor'. At Stanford, student occupations in 1969 convinced the university to sever its ties with the Stanford Research Institute (SRI), encouraging some research to cease and sending the remainder off-site.[4]

Moving out of the limelight, exploratory holographic research and military sponsorship became a more fugitive activity. Funding was increasingly focused, with the Defense Department expecting shorter-term military products and the National Science Foundation funding research conducted only at educational Institutions. University researchers found themselves increasingly channelled away from overt military applications of holograms, although covert, longer-term projects could continue under the umbrella of NASA funding. By the end of the decade, research on holograms had become less visible to the public and more carefully aligned to corporate goals.[5]

Diverse perspectives within the ranks of professional researchers were also evident. Interestingly, Dennis Gabor's books on science and society—a topic that had increasingly preoccupied him after promotion to Professor in 1960—unfashionably questioned the widespread confidence in technological progress and technocracy, a theme taken up independently by the student protesters and more overtly by Theodore Roszak, who documented their sometimes conflicting themes.[6] Competing notions about appropriate science and applications of holograms fostered distinct communities in specific milieus.

8.1 COUNTERCULTURAL EXPRESSIONS

The idea of holography was then, far more than now, imbued with a breathless sense of being on the threshold of some new form of consciousness.[7]

Randy James, 1979

[3] 'Labs' fate linked to ROTC', *Ann Arbor News*, 10 Dec 1969, 1; 'Willow Run future', *Detroit Daily News*, 4 Dec 1969, 2.

[4] Goodman, J. W., 'Research in Holography at Stanford: 1960s, 1970s and 1980s', unpublished report, SFJ collection, 30 Aug 2005; Goodman, J. W. to SFJ, email, 31 Aug 2005, SFJ collection.

[5] Gates, M., 'Holography a "re-creation of reality" – defense debate', *Ann Arbor News*, 18 Mar 1982, 2.

[6] Gabor, D., *Inventing the Future* (London: Secker & Warburg, 1963); Gabor, D., *The Mature Society* (London: Secker and Warburg, 1972).

[7] James, R., 'Off the wall', *Holosphere*, 8 (12), Dec 1979: 3.

The Bay Area was home not only to Stanford University and classified research on holograms, but also to the 1960s origins of student activism and a flowering of the counterculture in San Francisco.

A key contributor to countercultural holography was a Willow Run technologist and sometime PhD student working on lasers, Lloyd G. Cross. Cross had left in 1961 to found one of the first laser companies, Trion Instruments, and later worked on pulsed lasers for holograms at Kip Siegel's second company, KMS Industries. While there, his engineering interests took a new turn. He began to invite artists to the lab to see and talk about lasers and holograms. By 1968 he was developing lasers for light shows and public displays, patenting a device that would create a spot of laser light dancing in time with music. Cross showed off this new form of high-tech music to his peers at a local Optical Society of America meeting, and to the public at a local cinema and bars.[8] Light shows, lasers and holograms also successively attracted their contemporaries, many of whom were university students and young electronics hobbyists who identified similar possibilities. Among them were Richard Rallison, an Ohio State University student who founded a light show company, 'Electric Umbrella', in 1967, and Fred Unterseher, a light show operator and later a member of Cross's and Pethick's group.[9]

That year, Cross began to collaborate with artists, particularly Canadian Jerry Pethick (1935–2003), in creating laser art and holograms. Pethick devised a simple but effective optical table to keep their apparatus sufficiently still while recording holograms. With this 'sand table' and other clever low-cost innovations, he and Cross transformed holography from an expensive technology into a more affordable do-it-yourself activity.

With two other artists, Allyn Lite and Peter Van Riper, Cross and Pethick formed an Ann Arbor company, Editions Inc, to explore the possibilities of combining lasers, holograms, art and music. That first show allowed them to portray holograms not as products

[8] US Patent no. 3,590,861 filed 27 Nov 1968, granted 6 Jul 1971. A version of the product, dubbed *Sonovision* by Cross, was sold by Edmund Scientific from 1969, empowering a rapidly expanding wave of do-it-yourself light show enthusiasts and, soon afterward, would-be hologram makers.

[9] Holograms emerged alongside two other visual media of the 1960s, Op Art and light shows, and, like them, relied on visual surprise for its first impact. Op Art, explored by British artist Bridget Riley and others from 1960, employed stark black and white geometrical forms to generate confusing perceptions. Constructed from regularly varying lines and edges, they created visual sensations of depth, movement and even colour, but with an ever-shifting unstable quality. The images were meticulously executed with mathematical precision to yield a transcendent experience (Wade, N., 'Op art and visual perception', *Perception* 7 (1978): 21–46; Rycroft, S., 'The nature of Op Art: Bridget Riley and the art of nonrepresentation', *Environment and Planning D: Society and Space* 23 (2005): 351–71). Light shows, on the other hand, relied on time-dependent visual shock: strobe lights flashing at frequencies similar to those of the brain (occasionally leading to seizures in susceptible individuals), and vibrant and juxtaposed colours changing shapes with the rhythm of the music. Both these dissonant visual experiences generated visual confusion and disorientation. Like holograms, they became emblematic of particular cultural periods: the 'swinging [mid] sixties' for Op Art, after it was taken up by clothing designers and interior decorators, and the 'psychedelic' era of the late 1960s, when light shows briefly became the trademark of live performances of countercultural music groups such as Jefferson Airplane and Pink Floyd (Hewison, R., *Too Much: Art and Society in the Sixties 1960–1975* (London: Methuen, 1986)).

of big science, but as a new part of popular culture that would be delivered by innovative designers. Cross mused,

> Within a year or so, I think there will be hundreds of little holographic studios all over the country with people exploring holography the way photography was explored. Commercial holography is now where photography was in the mid-nineteenth century, and the next step will be to develop a simple, cheap pulsed laser—this will do for popular holography what the flash bulb did for photography.[10]

The enthusiasts mounted some of the earliest art-oriented hologram exhibitions and seeded public interest. At Detroit's Cranbrook Academy of Art in November 1969, *The Laser: Visual Applications* displayed their laser light show and growing portfolio of holograms. The exhibition encouraged staid audiences to lose their inhibitions in unfamiliar environments—crawling into an inflated plastic bubble, sitting on beanbag chairs and adapting to semi-darkness to peer at holograms—most of them for the first time. The holograms were 'more artistic' and clever than those that the Conductron engineers had imagined, ranging from disappearing pies to jigsaw puzzles.[11]

Cross himself—a product of WRL and a series of start-up electro-optical companies—felt carried along in a cultural current. 'I had finally found something that I wanted to put my time and energy into for a number of years', he recalled. 'It wasn't as much of a decision as a revelation to me. I could really feel my interest and my energy going into this craft, art, trade'.[12]

Over the next two years, Cross and Pethick produced light shows and hologram exhibits that toured theatres around New York State, culminating in *N-Dimensional Space* (Figure 8.2), the first hologram exhibition in New York City, but ultimately gravitated towards San Francisco.[13]

Ideas circulating on the West Coast resonated more closely with their notions of design and aesthetics. Buckminster Fuller (1895–1983), designer and futurist, had been promoting his unconventional concepts to American audiences from the 1930s, but found particularly enthusiastic converts among the baby boom generation. Fuller reached large audiences through his invention of the geodesic dome, the largest example of which was the American pavilion at the 1967 World's Fair in Montreal. Like holograms themselves, the design had first proven attractive to the American military during the 1950s for rapid-assembly huts, but was being enthusiastically adopted a decade later by members of counterculture communes. The structures embodied Fuller's notion of 'tensegrity', an engineering concept that allowed light, easily-fabricated and inexpensive shelters, but also chimed with countercultural themes such as frugality, adaptability and

[10] Wolff, M., 'The birth of holography: a new process creates an industry', *Innovations* (1969): 4–15.

[11] Charnetski, C. to SFJ, interview, 3 Sep 2003, Ann Arbor, MI, SFJ collection; Cranbrook Academy of Art, 'The Laser: Visual Applications,' Detroit, Michigan, 1969.

[12] Cross, L., 'The Story of Multiplex', transcription from audio recording, Naeve collection, Spring 1976.

[13] Finch College, Museum of Art, Contemporary Study Wing, T. McBurnett and E. H. Varian, 'N-dimensional space,' New York, 1970.

Fig. 8.2 Lloyd Cross with hologram (*N-Dimensional Space* exhibition catalogue (New York: Finch College Museum, 1970; uncredited photographer), courtesy of D. Broadbent).

self-sufficiency which were ideal for building holography equipment. Such principles were underlain by a philosophy of sustainability, which Fuller promoted in seminars, interviews and writings.[14]

A vehicle for such ideas and compendium of many more was *The Whole Earth Catalog*, published just south of San Francisco from 1968. As the most widely circulating periodical of the counterculture, the *Catalog* fell somewhere between underground newspaper and technical magazine. It was a compendium of book reviews, pragmatic designs, political invectives and philosophical texts, cut-and-pasted to inform, inspire and empower the counterculture generation.[15]

In San Francisco, Cross and Pethick earned money by offering courses from 1971 on holography to 'biker guys, to little old ladies, housewives and hippies'.[16] Their equipment was Fullersque and *Whole-Earth*-like, combining curved plywood sheets as room dividers, particle-board, inner tubes and PVC tubing for optical benches, and door springs as gears. Operating the following year as a technical commune in a rented and spacious hanger-like building, most participants and hangers-on at the School of Holography (SoH) were

[14] See, for example, Fuller, B., *Operating Manual for Spaceship Earth* (New York: E. P. Dutton & Co., 1968) and Fuller, B. and E. J. Applewhite, *Synergetics: Explorations in the Geometry of Thinking* (New York: Prentice Hall, 1982).

[15] Brand, S., *Whole Earth Catalog—Access to Tools* (Menlo Park, CA: Portola Institute, 1970). See also Turner, F., *From Counterculture to Cyberculture: Stewart Brand, the Whole Earth Network, and the Rise of Digital Utopianism* (Chicago: University of Chicago, 2006) and Kirk, A. G., *Counterculture Green: The Whole Earth Catalog and American Environmentalism* (Lawrence: University Press, Kansas, 2011).

[16] Unterseher, F. to S. F. Johnston, interview, 22 Jan 2003, Santa Clara, CA, SFJ collection.

drawn from their students. While short-lived and permeable, the San Francisco School's practices, values and products seeded a subculture, and its members experienced the interactions that anthropologists would identify as kinship. For its diverse contributors, this technological commune offered not just practical skills and access to a state-of-the-art technology for personal creativity, but a sense of common purpose. Just as importantly, they proselytized: their members wrote the first books dedicated to amateur holography,[17] and over the next few years, the School trained a generation of entrepreneurs and artists in the methods of making holograms.[18] A number applied their skills to commercial ends—creating and selling holograms in the Bay Area, and setting up similar schools in New York and, soon thereafter, in European cities.[19]

8.2 MEDITATING ON HOLOGRAMS

> *Looking at a hologram is not like looking at anything else. There's something totally different about it. I have some holograms that I've made, like two years ago, and I still, whenever I look at that hologram, I look at it for the first time, every time. It doesn't fade, in some strange way.*[20]
>
> Lloyd Cross, 1976

Countercultural expressions went beyond mere personal empowerment or the advantages of being early entrants in a new technological field. The members of the schools of holography of the 1970s did not map neatly onto the technical hobbyists of the late sixties. Instead of practising a tame version of modern science, the schools encouraged reinvention of science for new purposes. This social environment favoured holograms of new types, uses and symbolic meanings.

A few weeks before starting their classes, Cross and Pethick were interviewed on KFPA, Berkeley's student-oriented radio station, about the emerging field. Their new mixture of science and pop culture gently stretched the preconceptions of the interviewer, composer/poet Charles Amirkhanian, as it clarified their own thoughts:

INTERVIEWER: Why would anybody want to make a hologram?

PETHICK: 'Well, uh . . . I'm not really very certain. I think one of the reasons for making it is because you see something you've never seen before, and it has the possibility of, like, arriving at some place you don't know about. It doesn't go in any direction, because it's totally new. And because of that, it's a very exciting thing. I think the implications of it, and the applications, for communication in, like, a visual way, can be socially very

[17] Pethick, J., *On Holography and a Way to Make Holograms* (Burlington, Ontario: Belltower Enterprises, 1971); Dowbenko, G., *Homegrown Holography* (Garden City, NY: Amphoto, 1978); 6.

[18] Gorglione, N., 'Lloyd Cross', *Holographics International* 1 (1987): 17, 29; Albright, T., 'School of holography', *Radical Software* 2 (1972): 56–7.

[19] Burns Jr, J., 'Update on the New York School of Holography', *Holosphere*, 4 (4), Apr 1975: 3–4.

[20] Cross, L., 'The Story of Multiplex', transcription from audio recording, Naeve collection, Spring 1976.

strong. It's like, in a sense, like television was, it's eventually going to come that people can relate in a three-dimensional volume that has no substance of any kind . . .'

INTERVIEWER: Can you make a motion picture out of holograms?

PETHICK: There's been a couple of samples of that on a very small level. [But] you're applying the concept of film and moving still photography to the concept of holography which might be completely . . .

CROSS: Yeah, I like to think about that as a linear projection of the use of holography. We have, like, TV today, then 3-D TV; we have movies, are we going to have 3-D movies? I don't think those linear projections are necessarily going to be the most interesting thing that's going to happen. I think holography is such a powerful medium that the most interesting things are going to be totally unpredictable.

INTERVIEWER: Do you see applications for communication worldwide, say for international communication?

CROSS: Not so much through holography alone, but through the use of the laser, which is basically a high information rate carrier [. . .] whereas the hologram is for storing information.

INTERVIEWER: Would it be possible through the use of holography, to be sitting in Washington, say, and talking to somebody in Moscow and walking around to witness the back of his head and talk to him and look at the front of his face . . . is that possible through holography?

CROSS: There again, you can say that is definitely possible; the laser has the bandwidth to carry that kind of information, and holography has the storage capacity to store the whole image of a person. But that's one of those linear projections that may or may not come true, and if they do, they may be just nothing more than another telephone . . . I think there are other aspects of holography, which I can see indications of, that are very hard to predict, that I just have feelings about. It may never evolve into a totally moving medium itself, but it may evolve into a tremendous environmental medium. Holography is not just three-dimensional photography.[21]

Throughout the half-hour interview, backed by the ethereal sounds of the programme's signature 'space music', Cross and Pethick returned repeatedly to the 'perfect light' of lasers, the excitement of the new visual technology, and 'nonlinear projections' about the future possibilities of holograms as a meditative medium.[22]

[21] Cross, L. and J. Pethick, 'Lloyd Cross and Gerald Pethick on holography', *Ode to Gravity*, (KFPA, 20 Oct 1971), http://radiom.org/detail.php?omid=OTG.1971.10.20 retrieved 15 Jan 2013.

[22] Such futuristic soundscapes were a feature of several subsequent exhibitions, notably the *Light Dimensions* show at the Science Museum and *Light Fantastic* in Covent Garden, London, which looped a commissioned piece, 'Light Music', by Geoff Leach (Richardson, M., communication to S. Johnston, 9 Jan 2015).

One dimension gradually explored by the counterculture artisans was the consideration of holograms as sources of unearthly imagery. Lasers had kick-started this attention: reflections of laser beams off reflective surfaces produced unexpected patterns of light and dark—interference fringes and caustics of fascinating complexity. Holograms could not only record such patterns, but overlay them, too, in three-dimensions. And with the appropriate optical set-up it became possible to record white-light holograms that reproduced colours of any hue.[23] None of this was much valued or developed in the corporate labs, but became one of the most appealing aspects for the members of the counterculture technical community.

On an individualistic level, artisanal hologram-makers could explore visual perception, consciousness and transcendence. And technologies were combined promiscuously: psychotropics enhanced visual experiences. Recreational drugs could alter perception while viewing holograms; alternatively, the visual experience of such a trip could be evoked by creating appropriate holograms. Holographic artist and portraitist Ana Maria Nicholson recalled,

> Holography was a mission—we were here because these were images that came from the other side, as it were—it was a magical medium, and we thought that it would really transform everything; it affected me a great deal.[24]

Such explorations encouraged the creation of holograms that inspired reflection and meditation. These often abstract holograms shared little with the knick-knacks and lab equipment recorded in the holograms made by Ann Arbor engineers a few years earlier.

Like the forgotten arts of alchemy and astrology, making artisanal holograms reconnected the physical and symbolic dimensions of experience. Science and the conjurer's art were combined. *Rolling Stone* magazine captured the visual and metaphysical seduction of holograms for its readers:

> The hologram is as likely as anything technological to push your subliminal awe and wonder button and leave an ancient message flashing somewhere below the surface of consciousness: Here we have some Powerful Magic.[25]

Holograms were not merely scientific tools anymore; they had become countercultural mandalas ideal for meditation and replete with symbolic meanings.

These community explorations also revived and extended the metaphors that had accompanied the first holograms of the early sixties. Physicist David Bohm (1917–92) mused publicly about an analogy between holography, human perception, and physical reality.[26] Psychophysiologist Karl Pribram (b. 1919) similarly promoted the analogy of

[23] These synthesized colours could be generated by clever optical arrangements when recording rainbow holograms, a medium promoted by artists from the mid-1970s.

[24] Nicholson, A. M. to SFJ, interview, 21 Jan 2003, Santa Clara, CA, SFJ collection.

[25] 'Do-it-yourself holography', *Rolling Stone*, (18), 30 Aug 1973: 40.

[26] Bohm, D., 'Quantum theory as an indication of a new order in physics. II. Implicate and explicate order in physical law', *Foundations of Physics* 3 (1973): 139–68.

human memory as holographic. His 'holonomic' model suggested that electrical oscillations of the brain, acting on synapses, might produce effects similar to the interference of laser light in a hologram plate.[27] Both made much of the fact that any portion of a Leith–Upatnieks hologram contains a view of the entire three-dimensional scene. The hypotheses of Bohm and Pribram were frustratingly imprecise and untestable, but proved intriguing to non-scientific audiences. For the following countercultural generation, these links between holism and holography resonated even more closely with eastern theologies and mystical elements in counterculture thinking.[28]

8.3 MAKING HOLOGRAMS MOVE

> *We did holograms for Ripley's Believe it or Not, for Logan's Run and the Man Who Fell to Earth, everything you could think of. There was a brilliant idea, way ahead of its time, for a holographic pet-finding pedigree service . . . We felt it was all going to happen any minute.*[29]
>
> Fred Unterseher, Santa Clara, CA, 2003

Lloyd Cross was a transitional figure straddling the closed corporate structure of classified contracts and the open-source aesthetic of the counterculture.[30] To his students and associates, he was a guru who had rejected military applications for communitarian ideals. To other engineers, he was alternately a 'strange, hippy guy' or 'free spirit scientist'.[31] Cross could combine mind-expanding concepts with rhetoric heavily flavoured by his former employer at KMS Industries, Kip Siegel:

> Considering such things as holographic television, mass transference via laser beam, projection in free space without screens and stuff like that, either forget it forever as a totally fucked up idea or maybe wait ten or fifteen years, if we last that long, for some kind of holographic 3-D video.[32]

While Cross and Siegel's other engineers had in the past contrived demonstrations of full-colour holograms and limited animation to lure investors, the SoH pursued commercial products. Its methods were artisanal and its equipment improvised, but its principal product, the so-called multiplex hologram (known generically as an integral hologram or holographic stereogram), was far more successful than anything produced by Siegel's companies.

[27] Pribram, K. H., 'The neurophysiology of remembering', *Scientific American* 220 (1969): 73–86.

[28] See, for example, Wilber, K., *The Holographic Paradigm and Other Paradoxes: Exploring the Leading Edge of Science* (Boulder, CO: Shambhala, 1982).

[29] Unterseher, F. to S. F. Johnston, interview, 22 Jan 2003, Santa Clara, CA, SFJ collection.

[30] Turner, F., From Counterculture to Cyberculture: Stewart Brand, the Whole Earth Network, and the Rise of Digital Utopianism (Chicago: University of Chicago, 2006); Braunstein, P. and M. W. Doyle, Imagine Nation: the American counterculture of the 1960s and '70s (London: Routledge, 2002).

[31] Haines, K. to SFJ, interview, 21 Jan 2003, Santa Clara, CA, SFJ collection; Leith, E. N. to SFJ, interview, 22 Jan 2003, Santa Clara, CA, SFJ collection.

[32] Cross, L., 'The potential impact of the laser on the video medium', *Radical Software* 1 (1970): 6.

Cross and Pethick had tried to steer their 1971 radio interviewer away from 'linear predictions', but Cross did muse about practical ways of producing moving holograms. The members of the SoH developed and sold a novel form of hologram that, for the first time, brought animated images to wide audiences. Cross's 'multiplex hologram' combined many holograms into a single view, thereby 'multiplexing' them.[33] Such ideas had been circulating amongst engineers from the late sixties, but Cross developed a number of appealing innovations. His hologram was a 44 × 9 inch filmstrip curved to form a 14-inch diameter cylinder. The hologram was in fact a series of narrow vertical strip holograms, each recorded from a single frame of movie film. The original movie film had recorded a subject on a rotating platform, or an outdoor view in which the camera recorded a moving scene. The integral hologram was then recorded strip by strip using a sophisticated hologram printer improvised by assembling a laser, film projector and repurposed components. An additional innovation was to incorporate an optical arrangement that would make the resulting hologram viewable in white light. The result was an animated rainbow hologram viewable from all sides (Box 8.1).

When the resulting hologram was illuminated by an unfrosted light bulb from below, it reconstructed an image in the middle of the cylinder. As observers walked around it, their eyes intercepted different 'frames' from any viewing position, creating a visual effect equivalent to a stereoscopic movie lasting some 10 to 20 seconds. The multiplex hologram thus provided a fourth dimension—time—and unlimited subject matter, because the original

Box 8.1 Viewing integral holograms

Integral holograms (also known as 'Multiplex' holograms or 'holographic stereograms') consist of a series of vertical strip holograms, recorded and arranged as either a cylinder (as in Cross holograms) or as a flat plate (as in the works of artist Patrick Boyd and others) and lit by a compact white-light source. As the viewer moves sideways—or as the hologram rotates—the eyes intercept different strip holograms. The experience usually reveals a three-dimensional scene and a few seconds of animation. The scene itself may be as varied as those for Victorian stereoscopes, and can be recorded with a movie camera or from synthesized digital imagery.

Integral holograms are often recorded as white-light transmission (rainbow, or Benton) holograms. As the viewer moves up and down, the colour of the image will shift across the spectrum, and there will be no change of vertical perspective. They may also be created as laser-lit (Leith–Upatnieks) transmission holograms or as Denisyuk reflection holograms.

[33] The term multiplex was first used to describe the encoding of multiple communications streams onto a single carrier transmission. Like many similar terms, it merged technical cultures by linking the theory of electronic communications to modern optics.

subject was recorded with a cine camera rather than a laser.[34] The spectacle was reinvented: for the first time, holograms could record three-dimensional moving scenes outside the lab, and they could be replayed without a laser or special viewing glasses (Figure 8.3).

In their Bay Area workshop, the members collectively printed, chemically processed and packaged tens of thousands of their holograms. The proof-of-concept image of Cross's girlfriend breaking into a smile and blowing a kiss was the prototype of the Multiplex Company's most popular hologram, *Kiss II*. That hologram—in both a life-size 19-inch cylinder and a more portable six-inch version—became a frequent display at science centres and galleries during the late 1970s, and iconic of the new holography.

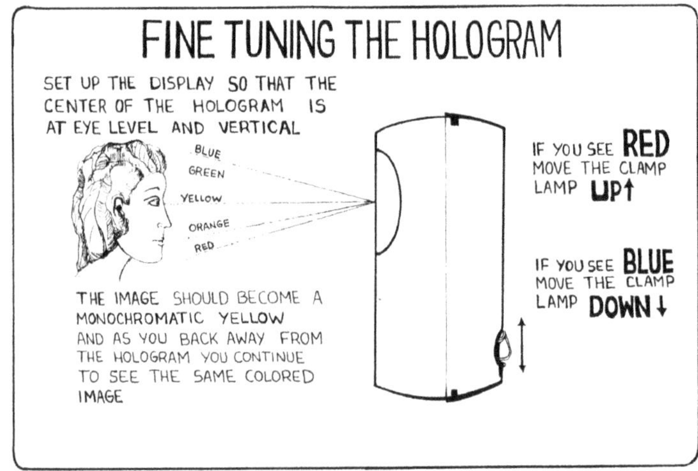

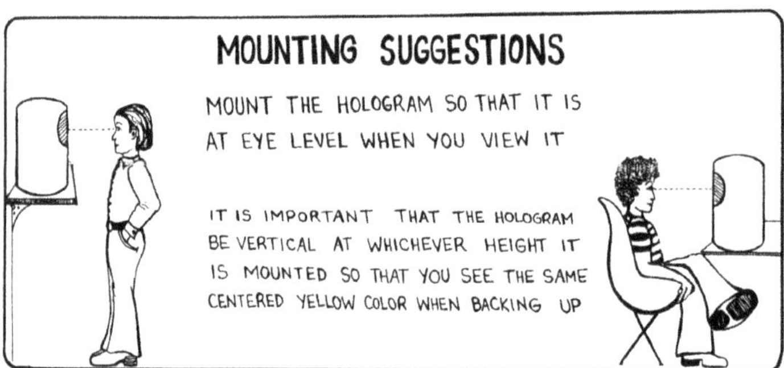

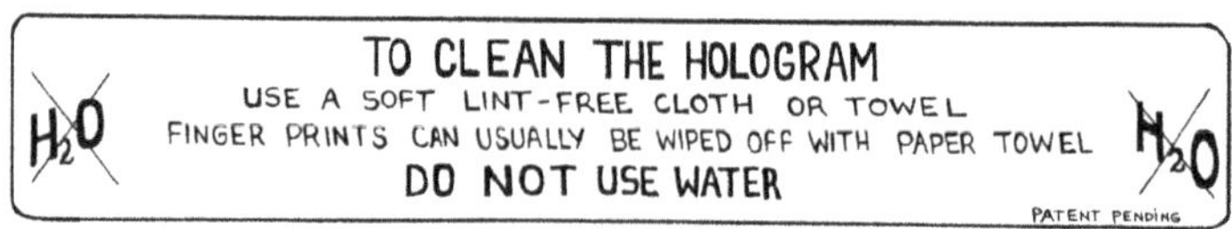

Fig. 8.3 How to view a hologram, 1970s style (Multiplex Company, c1975; S. Johnston collection).

34 Schroter, J., 'Technologies beyond the still and the moving image: the case of the multiplex hologram', *History of Photography* 35 (2011): 23–32.

The countercultural origins led increasingly to conventional commercial aims. The company's affordable products encouraged public display and promoted direct purchases by consumers—an exemplar for a cottage industry for holograms by the late 1970s. The company capitalized on its product by filming dozens of subminute movies for the Multiplex treatment. They sought to attract varied audiences with imagery of dance moves, rotating molecules, tourist views of cable cars, Kung-Fu fighting and scenes from horror films (Figure 8.4). Edmund Scientific became an important outlet, and science centres and technology museums in a growing number of cities stocked the holograms and their simple display units. As with stereo views a century earlier, private viewers were also cultivated: suiting the relatively liberal 1970s, a significant fraction were of mildly erotic subjects. For more particular tastes, under-the-counter pornographic holograms were also available. Among the most widely seen was *Pam and Helen*, which cultural theorist Umberto Eco saw as emblematic of modernity itself:

> Two very beautiful naked girls are crouched facing each other. They touch each other sensually, they kiss each other's breasts lightly, with the tip of the tongue. They are enclosed in a kind of cylinder of transparent plastic. Even someone who is not a professional voyeur is tempted to circle the cylinder in order to see the girls from behind, in profile, from the other side. The next temptation is to approach the cylinder, which stands on a little column and is only a few inches in diameter, in order to look down from above: but the girls are no longer there . . . [35]

Fig. 8.4 Viewing a Multiplex animated hologram, 1975 (Berner J., *The Holography Book* (New York: Avon Books, 1980), illustration p. 38 by Steven L. Borns with permission of J. Berner). Compare with Figure 3.8.

[35] Eco, U. and W. Weaver, *Travels in Hyper-Reality: Essays* (London: Picador, 1986), pp 3–4. See also Johnston, S. F., *Holographic Visions: A History of New Science* (Oxford: Oxford University Press, 2006), pp 396–7.

By the late 1970s, three multiplex printers were in operation, in San Francisco, Chicago and New York.[36] Distributors sought new markets for stock and custom-made multiplexes. White Light Works (Woodland Hills, CA), for example, marketed them as 'People Stoppers', touting their use in advertising:

> RCA Records . . . in record stores to introduce a new recording artist
> ARCOA . . . as a traffic gatherer in their annual trade show
> Magic Mountain . . . as a point-of-purchase display to help sell tickets for their amusement park
> Millers Outpost . . . for an attention getter at grand openings
> American Medical International . . . as a trade show People Stopper to introduce a service to the medical field
> Microsound Communications Corporation . . . in a training convention to display operation of a new communication device[37]

For a cost of several thousand dollars, affluent buyers could even have a custom Multiplex hologram recorded of themselves, with copies for a couple of hundred dollars more—the first holographic portraits available outside laser laboratories.[38] By the end of the decade, other innovators were marketing improved versions of the technique to reduce its distortions and to achieve the same animation from a flat hologram.

8.4 HOLOGRAMS AT SCHOOL

The members and former students of the holography schools of the 1970s became more accessible spokespersons for the field than the scientists and engineers before them had been. They supplemented their incomes and popularized their craft through classes, how-to books and holograms themselves.[39]

Such countercultural products produced waves of public attention. Multiplex holograms, along with the flat alternatives in the form of image-plane, reflection and rainbow varieties, could be displayed in any lobby or shop that had a spare electric socket for a light source. Average people—not just engineers and amateur scientists—could create their own. The SoH brought ubiquity and empowerment to holograms.

These changes fostered new communities, turning holograms from a solitary interest into a shared enthusiasm. Conventional educational channels increasingly paralleled the courses available in countercultural environments. As Edmund Scientific discovered, teachers could marshal adequate budgets and maintain classroom enthusiasms.

[36] Ross, J. to SFJ, interview, 3 Apr 2003, London, SFJ collection.
[37] White Light Works Inc, brochure (Woodland Hills, CA, c1980); G. Zellerbach collection.
[38] Multiplex Company catalog (San Francisco, 1978); G. Zellerbach collection.
[39] Notably Pethick, J., *On Holography and a Way to Make Holograms* (Burlington, ON: Belltower Enterprises, 1971); Dowbenko, G., *Homegrown Holography* (Garden City, NY: Amphoto, 1978); Unterseher, F., J. Hansen and B. Schlesinger, *Holography Handbook: Making Holograms the Easy Way* (Berkeley, CA: Ross Books, 1981).

Holograms proved to be an effective conduit from countercultural do-it-yourself activities into teacher-led high school projects.

But—as with technical amateurism and the cultural appeals of holograms—educational uses had earlier roots. The origins of classroom exposure extended back to the earliest research on holograms and piggy-backed on Cold War initiatives to boost national science.[40] Physicist Albert Baez (1912–2007), while a PhD student at Stanford University during the late 1940s, had reproduced and extended the hologram experiments of Dennis Gabor. His background as a technical amateur was typical of his times. Baez recalled with pleasure 'the fascination that accompanied the first experience' of processing a roll of photographic film, and had joined a radio club in his manual training high school:

> I built a very primitive television receiver in 1928. It was a mechanical device that looked nothing like a modern TV. It used a circular scanning disk which had to rotate at the same speed as the disk at the broadcasting station. That problem had not been solved, so I wrote a letter to the editor of the magazine *Radio News* suggesting how it might be done. The editor, Hugo Gernsback . . . wrote me a nice letter saying that they would print it and that I would receive a free one-year subscription. I now consider these events my first invention and my first publication.[41]

The frugality inculcated during the Second World War was an important component for his generation of innovators. Baez recalled that because Stanford's physics department had been starved of resources his supervisor, Paul Kirkpatrick, had told him to 'go the scrap metal pile in the machine shop and see what pieces we can utilize in our preliminary experiments'.[42]

When he took a teaching post at Redlands University in 1950, Baez wanted to improve the educational opportunities of his students. Kirkpatrick suggested that Gabor's experiments could be an ideal activity for undergraduates. By using available equipment such as mercury arc lamps and photographic film to record and view holograms, they would gain direct experience of research. With financial support from Research Corporation, Baez led successive classes through the process. Their experiments sought to 'combine pleasure and knowledge' and, 'to induce them to linger', students were 'allowed to play phonograph records'.[43] Baez employed a five-student undergraduate research group one summer to study the technique further (Figure 8.5). Trained between 1952 and 1956, this small cohort was the first generation to explore

[40] Rudolphs, J. L., *Scientists in the Classroom: The Cold War Reconstruction of American Science Education* (New York: Palgrave, 2002).

[41] Baez, A. V., 'The early days of x-ray optics – x-ray microscopes, telescopes and holograms,' invited talk, Lawrence Livermore National Laboratory, CA, 1999, typescript, S. Johnston collection; quotation p.1; Reimers, F., 'Albert Vinicio Baez and the promotion of science education in the developing world 1912–2007', *Prospects* 37 (2007): 369–81.

[42] Baez, A. V., 'Pinhole-camera experiment for the introductory physics course', *American Journal of Physics* 25 (1957): 636–8, quotation p. 636.

[43] Baez, A. V., 'Some notes on laboratory procedures in a small college', *American Journal of Physics* 25 (1957): 386–7, quotation p. 386.

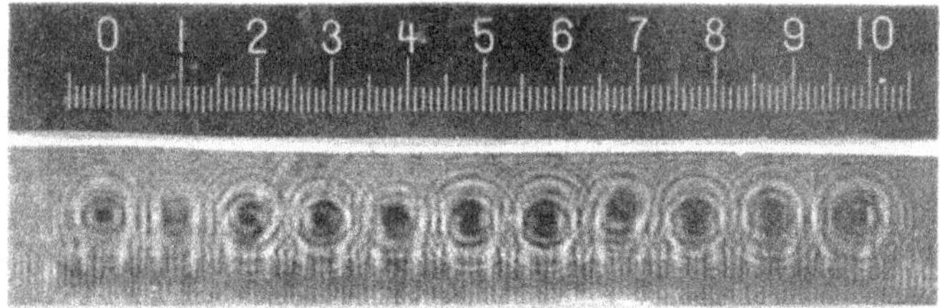

Fig. 8.5 Hologram by Albert Baez and his students, c1954. Top: original object (a microscope reticule); bottom: corresponding in-line (Gabor) hologram (Baez, A. 'The early days of x-ray optics (presentation slides)' (1999, unpublished). The donations of Albert Baez documents by Skip Henderson are gratefully acknowledged).

the principles of wavefront reconstruction, and the only non-professional audiences of the era to see holograms.[44]

As with the earlier technologies of stereoscopes, films and radio, a handful of educators of the era recognized the teaching potential of technological demonstrations. From the 1960s, van de Graaff generators, liquid nitrogen and helium balloons became ubiquitous in high school science labs for demonstrating scientific principles. The invention of holograms—and the lasers that produced them—appeared at the peak of this wave. Holograms could be a means of engaging students in the excitement of new science.

Albert Baez was unique in introducing holograms to classes during the 1950s, but a decade later, more impressive holograms were being widely publicized but still seldom seen. As individual adult enthusiasts were discovering for themselves, holography was an expensive and often disappointing pastime. After the publication of the hologram in the 1967 *Science Year Book*, the author of the accompanying article, Gary Cochran, had received a sprinkling of letters from high school students about science fair projects, but he discouraged them from trying to make a hologram.[45]

Viewing and making holograms in schools and colleges was within reach, though. Groups of students sharing the institution's equipment could create holograms that were still out of reach of individual amateurs. The second American academic to use holograms in the undergraduate lab was probably Tung Hon Jeong (1931–2015), a professor of physics at Lake Forest College, a small liberal arts institution in Illinois. From 1966, he began exploring the technique with available university lab equipment. Jeong also fused

[44] Baez taught an optics course for advanced undergraduates in autumn 1952 and 1953, and spring 1955 and 1956 (Gonzales, N. to S. Johnston, email, 19 Nov 2013, SFJ collection). See also Baez, A. V., 'Anecdotes about the early days of x-ray optics', *Journal of X-Ray Science and Technology* 7 (1997): 90–7; Baez, A. V. and G. Castro, 'A laboratory demonstration of the three-dimensional nature of in-line holography', *American Journal of Physics* 67 (1999): 876–9; Baez, A. V. to SFJ, email, 13 Mar 2003, SFJ collection. Like Kirkpatrick, Baez was an exponent of the educational role, as well as the research ambitions, of universities. He promoted science teaching through stints with UNESCO and focused on education later in his career, co-editing the book *The Environment and Science and Technology Education* (New York: Pergamon Press, 1987).

[45] Cochran, G. D. to SFJ, interview, 6 and 8 Sep 2003, Ann Arbor, MI, SFJ collection.

educational, academic and commercial activities, initially paralleling and extending what Albert Baez had begun a decade earlier.

Sponsored by the Gaertner Scientific Company, he wrote a 1968 classroom holography manual based on the lab equipment that the company manufactured for schools. The following year, Baez himself produced a $10 'holography teaching unit' aimed at primary school children, and in 1972, as the San Francisco SoH was getting off the ground, Jeong and Baez collaborated to make a classroom instructional film on holography for the Encyclopaedia Britannica Educational Corporation. Jeong formed his own company, Integraf, in 1973 to distribute experimental and instructional materials and educational kits. He rapidly dominated this niche, offering short summer courses at the College for students, technologists and interested laypeople. Jeong's networking allowed him to collate a growing database of persons active in the field. The result of these activities was a holistic glimpse of a field that was still a fragmented and esoteric art.[46]

Such guidance permitted a first generation of science fair projects during the 1970s, made feasible by the increasing availability of school-owned lasers (Figure 8.6). Jeong's subsequent organization of conferences on optics and holography, including a series at

Fig. 8.6 British school holography project, 1986 ('Leicestershire holography project', *Holosphere* 15 (2) (1987): 22; J. Bellamy photo; courtesy of MIT Museum).

[46] Jeong, T. H., *Gaertner–Jeong Holography Manual* (Chicago: Gaertner Scientific Corp., 1968); Baez, A. V. and T. H. Jeong, 'Introduction to Holography', (Encyclopaedia Britannica Educational Corp., 1972); Jeong, T. H., *A Study Guide on Holography (Draft)* (Lake Forest, IL: Lake Forest College, 1975); Jeong, T. H., *A Demonstrated Lecture on Holography* (Chicago: Instant Replay, 1978); Jeong, T. H. and F. E. Lodge, *Holography using a Helium–Neon Laser* (Blackwood, NJ: Metrologic Instruments, 1980).

his institution (the 'International Symposia for Display Holography' from 1982), attracted a growing variety of enthusiasts, artists, educators and professionals to the expanding field.[47]

Holograms found a more frequent place at college level. Much as Baez and Jeong had pioneered, undergraduate physics lab courses around the world increasingly included a hologram experiment from the early 1970s as helium-neon lasers became a ubiquitous piece of teaching apparatus. For entrants with bachelor's degrees, there were numerous physics departments offering research degrees relating to holography; from the late 1960s, the number of dissertations averaged over twenty-five per year drawn to established research clusters at University of Michigan, Stanford, Imperial College, London, and others.[48] A later centre for research on display holograms was Stephen Benton's contribution to the Media Arts and Sciences programme at the MIT Media Lab, 1987–2003. Also at postgraduate level, a handful of art colleges included instruction on hologram-making. The most influential of these—producing a generation of British art holographers—was at the Royal College of Art in London, which operated a Holography Unit in its Photography Department between 1985 and 1994. During the late 1980s it was the only institution offering masters and doctoral degrees in art holography and over its lifetime trained some thirty graduates.[49]

Other art colleges offering training in holography included MIT's Center for Advanced Visual Studies; the Ontario College of Art Holography Department; the School of the Art Institute of Chicago, offering Bachelors and Masters degrees in Fine Arts; the University of Tsukuba, Tokyo, which included holography in its Plastic Arts and Mixed Media Course from the 1980s; the Academy of Media Arts Cologne, Germany; and, in Australia, holography laboratories at the Sydney College of the Arts, the Design School of the University of Technology the College of Arts of the University of New South Wales.

8.5 ENTHUSIASTS' NETWORKS

As opportunities for seeing and making holograms grew, popular interests consolidated. Individuals exposed to the short courses on holography or experiments in high school and university exploited their new skills and enthusiasms as competent hobbyists,

[47] Jeong, T. H., P. Rudolf and J. Luckett, '123 360 degree holography', *Journal of the Optical Society of America* 56 (1966): 1263; Jeong, T. H., 'Holography comes out of the cellar', *Optical Spectra*, 2 (6), Jun 1968: 58–65; Jeong, T. H., 'Holography and education in the United States', presented at *Three-Dimensional Holography: Science, Culture, Education*, Kiev, USSR, 1989; Jeong, T. H., 'Making holograms in middle and high schools', in *Proceedings of the SPIE Sixth International Conference on Education and Training in Optics and Photonics*, Cancun, Mexico, 2000; Jeong, T. H. to S. F. Johnston, interview, 21 Jan 2003, Santa Clara, CA, SFJ collection.

[48] Johnston, S. F., *Holographic Visions: A History of New Science* (Oxford: Oxford University Press, 2006), Appendix.

[49] Lloyd, S., 'The Holography Unit at the Royal College of Art', *Holosphere*, 14 (3), Summer 1986: 15–6; Murray, R., 'Holography at the Royal College of Art, London', *International Symposium on Display Holography* 1600 (1991): 237–9.

experimenters, artists, cottage business producers and consultants. Sharing two or more of these roles was common for participants.

Educational possibilities expanded during the 1970s and 80s, a period of rapid growth in public exposure to holography. The Lake Forest summer school was paralleled by others that lacked university or high-school connections, and often emerging from the model of the San Francisco SoH. The earliest of these were further education and recreational courses that attracted diverse audiences (Section 8.1), and later courses offered in colleges of art.[50] These varied initiatives sometimes faltered during the 1990s, when public interest in holograms was waning and students increasingly sought vocational skills (Figure 8.7).[51]

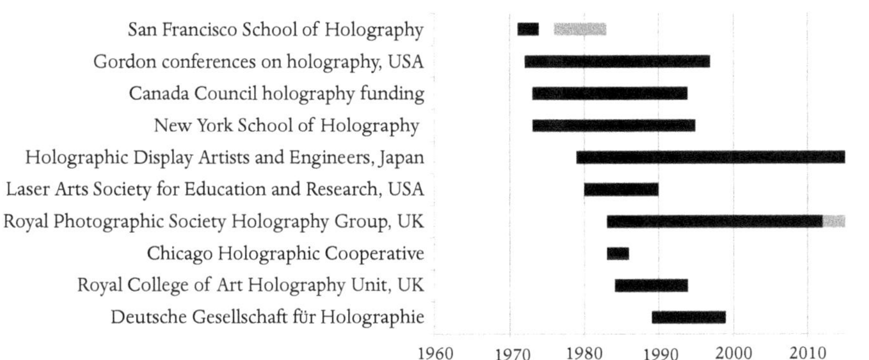

Fig. 8.7 Social and educational initiatives. Grey bars indicate scaled-down or reoriented activities.

[50] Stowman, W. P., *A holography laboratory for high school physics*, thesis, (1972); Robinson, M. L. A. and A. P. Saunders, 'Making holograms at school', *School Science Review* 55 (1973): 260–73; Johnson, G. T., *A holography unit for a junior high electronics laboratory*, thesis, (1983); Eichert, E. S. and A. H. Frey, *Holography in Driver Education, Training, Testing, and Research* (Springfield, VA: National Highway Traffic Safety Administration, 1978); Robbins, L. K., *Holography, a new medium for the high school art curriculum*, MA thesis, Ball State University (1980); Latham, R. E., 'Holography in the science classroom', *Physics Teacher* 24 (1986): 395–400; Cowles, S., 'The New York experience', *Holographics International* Winter 1988 (1988): 14–5; Jeske, M. A., *A high school optics unit emphasizing laser experiments and student production of various hologram types*, thesis, (1994).

[51] Jeong, T. H. and V. B. Markov, *Three-dimensional holography—science, culture, education: the international UNESCO seminar: 5–8 September 1989, Kiev, USSR* (Bellingham, WA: SPIE, 1991); Larkin, A. I., 'Holography and education in the USSR', *International Symposium on Display Holography* 1600 (1991): 412–7; Tomaszkiewicz, F., 'Continuing laser-imaging program of instruction in a public middle school', presented at *International Symposium on Display Holography*, Lake Forest, IL, 1991; Geelen, P., 'Holography in education: it's time to come out of the dark', *The Australian Science Teachers' Journal* 40 (1994): 15; Jeske, M. A., *A high school optics unit emphasizing laser experiments and student production of various hologram types*, thesis (1994); Pepper, A., 'A single beam recording system for creative holography and education', *Proceedings of the SPIE* 2043 (1994): 126–9; Tomaszkiewicz, F., 'Holography in public middle schools: perspectives on student safety and environmental impact', presented at *Fifth International Symposium on Display Holography*, Lake Forest, IL, 1994; Tomaszkieweicz, F., 'Ten-year perspective on display holography in the middle school classroom', presented at *Sixth International Symposium on Display Holography*, Lake Forest, IL, 1997; Connors, B. A., 'Lightforest and the MIT Museum Holography Education Project', presented at *Sixth International Symposium on Display Holography*, Lake Forest, IL, 1997; John, P., 'Teaching holography – inspiring an interest in science', *The Holographer*, http://holographer.org/media/articles/hg00003.pdf, 11 Apr 2004.

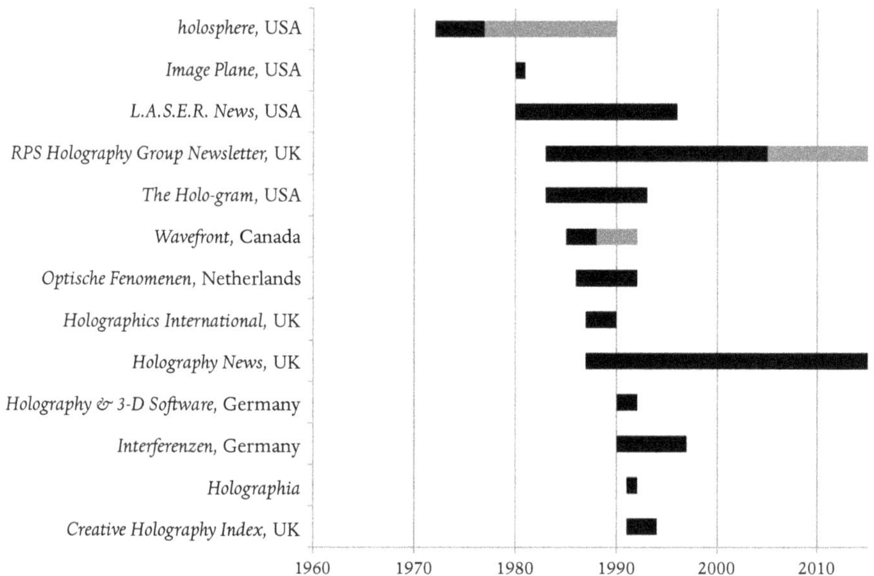

Fig. 8.8 Print periodicals dedicated to holography. Grey bars indicate reoriented or scaled-down activities.

Alongside these public waves came periodicals to carry a sense of community and shared knowledge further afield (Figure 8.8). Newsletters and magazines flourished during the 1980s, satisfying audiences unable to obtain or read scientific journals. The editors of the publications mirrored the attendees at the schools of holography, Lake Forest events and exhibitions. Some were individual enthusiasts promoting their passion; others sought to build communities of creators, consumers and critics.

From the late 1990s, the availability of semiconductor lasers made holograms less expensive to produce. Early models, developed via research on light-emitting diodes for fibre-optic communications, were not ideal for making holograms, but rapidly improved in specification and affordability. One enthusiast, Frank DeFreitas in Illinois, published a practical book that proved popular as an affordable entry into hologram-making.[52]

An equally important technological change was the popular uptake of the internet. For the first time, individual enthusiasts—who, by the 1990s, again dominated non-professional holography—could share their interests and skills. DeFreitas and Jeong were among the first to promote 'laser pointer holography' via websites aimed at, or created by, enthusiasts.[53]

[52] DeFreitas, F., A. Rhody and S. W. Michael, *Shoebox Holography: A Step-By-Step Guide to Making Holograms using Inexpensive Semiconductor Diode Lasers* (Berkeley, CA: Ross Books, 2000).

[53] Hologram-making kits—some eventually employing film that did not require traditional chemical processing, or supplied three laser diodes to produce colour holograms—were offered by companies such as Keystone Scientific, Integraf and Litiholo. Such make-it-yourself kits repeated the marketing approaches of Heathkit and Edmund Scientific two generations earlier.

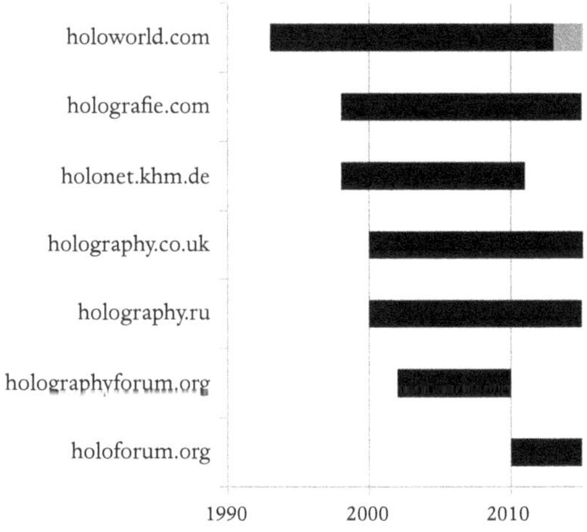

Fig. 8.9 Popular websites dedicated to holography. Grey bars indicate reoriented or scaled-down activities.

By the early 2000s, then, laser diodes liberated the viewing and making of holograms. Schools and short courses were no longer essential to indoctrinate and mentor enthusiasts. With the aid of a guidebook and a handful of components, individuals could be genuinely independent. But the growth of the internet during the same period encouraged a new kind of engagement: peer-taught holography via virtual communities (Figure 8.9). Like amateur radio enthusiasts and photographers, hologram-makers went online to collectively explore their pastime. Hologram enthusiasts, working largely alone but supported by a palpably growing network, began to rival professionals in their products and visibility.

9

Holograms on Display

From the mid-1970s, hologram cultures subdivided and blended in new ways. The products of artisanal cottage industry mutated into artistic creations. Countercultural themes were recuperated by cultural commentators and art critics. Museums and galleries sought new audiences by combining these representations of modernity with more traditional artefacts of cultural heritage. And a new wave of entrepreneurs sought to capture the zeitgeist in profitable merchandise. Established now as a visual medium, the hologram broadened its audiences. This new engagement with holograms probed questions at the heart of contemporary culture including the value of authenticity, the aesthetics of artistic expression and the appropriate role for modern science. The first step was towards the art world.

9.1 EXHIBITING HOLOGRAMS

> *My first memorable experience of holograms was at the Burnaby Art Gallery in 1976, where artist Al Razutis' Visual Alchemy exhibition displayed large reflection holograms lit by white light in darkened gallery spaces. The size and unfamiliar type of hologram gave them impact, but so, too, did their combinations. One work, 'Newtonian Galactic Assembly Line', was a yellow-green mosaic over a metre across (Figure 9.1). Others were mixed media pieces that combined reflection holograms with other objects. The unfamiliar constructions in the sparse and unpopulated gallery encouraged exploration and examination, and seeded the desire to make holograms.*
>
> Burnaby, British Columbia, 1976

Science labs had been the first venues for public viewings of holograms. The busy environments—physics labs humming equipment and queues of visitors—vied for viewers' attention. Just as interesting as the small image behind the glass plate was the helium-neon laser that lit the hologram, producing intense red light and shimmering granular reflections off the equipment.

The second wave was in a handful of small galleries. As pioneered with Lloyd Cross's Ann Arbor showings, the most successful of these exhibitions were produced by loose coalitions of technologists and artists. The constellation of interests offered something

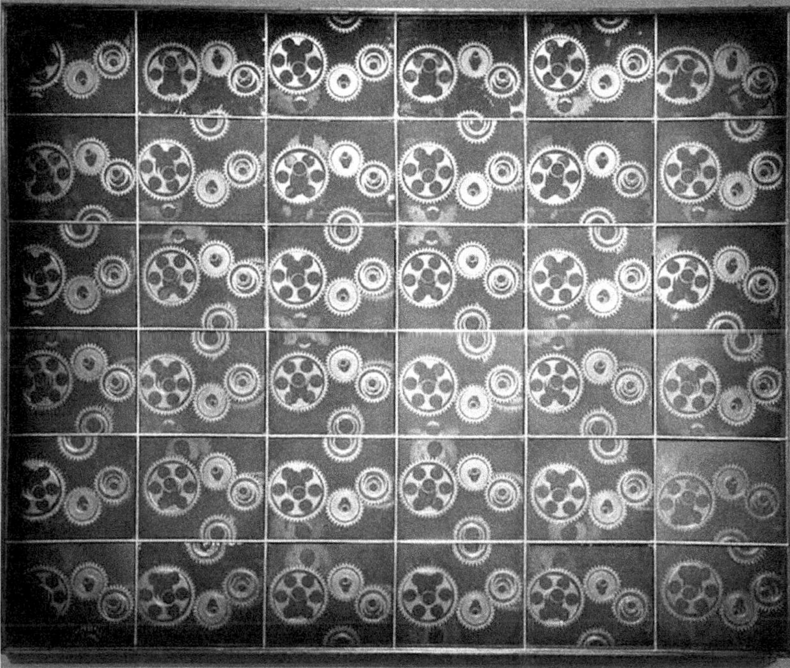

Fig. 9.1 'Newtonian Galactic Assembly Line', Al Razutis, 1976. (Silver halide on glass, reflection (Denisyuk) hologram mosaic 1.2×1.5 m^2. Photo courtesy of A. Razutis, Visual Alchemy).

for every visitor: high technology, magic and art, wrapped up in the still novel experience of multi-media.[1]

But the earliest hologram exhibitions made small ripples. Despite the intense memories of organizers and participants, the shows generated diffident interest from local entertainment reporters, producing few column-inches in newspapers and short items on local television news. Viewers' recollections are even more rare.

Early audiences were bemused, unprepared for how to view holograms and at times unaware even that a perceptual trick was being played. This was not a problem unique to holograms: as traced in earlier chapters, consumers have often grappled with new technologies, although this resistance is seldom noticed for long. Indeed, twentieth-century audiences have been so expectant of progress that user difficulties are often attributed to age or stubbornness.[2]

Public hologram displays initially proved both a technical success and consumer failure. An early example was Conductron's creation of a hologram display for General

[1] While lasers and holograms were mingling in late-60s exhibitions, cinema was being expanded by split-screen techniques and fine art was exploring a revival of mixed-media works.

[2] Corn, J. J., *User Unfriendly: Consumer Struggles with Personal Technologies from Clocks and Sewing Machines to Cars and Computers* (Baltimore: Johns Hopkins, 2011); see also Marvin, C., *When Old Technologies Were New: Thinking About Electric Communication in the Late Nineteenth Century* (Oxford: Oxford University Press, 1988).

Motors' New York headquarters in 1968. Representing the state of the holographic art, the display consisted of large 45× 60 cm transmission holograms in a box arrangement, allowing the image to be viewed from all sides. Each hologram had been recorded with two distinct images that could be reconstructed by a laser beam at the appropriate angle, thereby incorporating the unexpected dimension of time. Using sequenced laser lighting, the contents of the hologram cube showed the three-dimensional image of a nineteenth-century carriage magically transformed into a modern car body and—with a final change of lighting—revealed an empty box.[3] But to the chagrin of the engineers, most passers-by did not pause long enough to realize that the arrangement was anything more than a glass case displaying a model car. Thus 'real 3-D' had achieved a remarkable level of veracity, but could not compete with the real thing. As stereoscope views and widescreen cinema had demonstrated, holograms had to be both spectacular and fresh. A novel variety of visual realism was needed that nevertheless highlighted the role of the optical device itself.

Attracting and engaging visitors at hologram exhibitions consequently proved difficult. While evoking the familiar cultural appeals of magic and visual media, holograms evaded categorization. A reviewer of the first New York show, *N-Dimensional Space*, traced the scientific axis:

> It's neither a photograph, a movie, nor a 3-D projection. It's a hologram—art and technology's youngest and most provocative progeny . . . Holograms are still in their infancy, but these experiments indicate that a new and exciting visual medium is with us, offering limitless creative horizons.[4]

The first audiences at hologram exhibitions were attracted as much by lasers as by the holograms themselves. Cross and Pethick, with their shows in Detroit and New York State, promiscuously mingled holograms with dancing laser patterns and music. Indeed, until the early 1970s, laser-lit holograms (the type popularized by Leith and Upatnieks) were the easiest to make and by far the most common. But, as pioneering artist Margaret Benyon (b. 1940) discovered, this limited the works that she could display and shaped users' experiences. Forced to use monochromatic lamps instead of expensive lasers, her startling imagery was diluted: 'I had one laser that I'd borrowed from the lab, and a filtered mercury arc on another table, and a sodium lamp on one or two others, so they got progressively fuzzier'.[5] Familiarizing visitors also took unexpected effort. As earlier audiences had encountered the new visual grammar of cinema and graphic arts, they were unprepared for just what hologram-viewing involved. Benyon recalled of her first public exhibition of holograms,

> You've got this little Perspex thing here, edge lit, saying, 'please press button to turn on next hologram', and then arrows on the floor to show where they should walk, because no-one had ever seen one . . . And unfortunately the holograms were at table level as well, so people

[3] Wilfong, J., 'Hologram product display announced', *Conductron Antenna*, 2 Dec 1968: 4.

[4] Gruen, J., 'The whole message', *New York*, 3 (19), 11 May 1970: 59.

[5] Benyon, M. to SFJ, interview, 21 Jan 2003, Santa Clara, CA, SFJ collection.

> had to bend down to catch first sight of the images. Most people never got to see the holograms unless I was there to show them, even though I had put arrows and so on, because they didn't know what to expect; they had no idea. I had to teach them how to see the holograms.[6]

Such exhibitions—dependent on the seduction and availability of scarce lasers—were also self-limiting in another respect: lasers had developed a reputation for danger, too. The new technology, barely a decade old, had been the subject of demonstrations, news items and documentary films that showed off their unusual properties and potential power. The iconic image of a pulsed laser piercing a razor blade, or of an invisible carbon dioxide laser charring a black spot in wood, was memorable even if neither type was associated with holograms. And, as with other popular perceptions of holograms, earlier cultural experiences prepared the ground: lasers seemed to be the death rays forecast by interwar science fiction.[7] Lloyd Cross and his associates reassured visitors to their show that 'laser beams of the low power used will not burn or harm human flesh in any way', but the impressions left by *Goldfinger* were hard to forget.[8]

During the 1970s, though, the availability of new types of white-light viewable holograms liberated exhibitions from the perceived dangers of laser beams. Scientists and engineers stuck with laser-lit holograms because of their faithful reproduction, but the schools of holography learned that white-light holograms were often as easy to make and much easier to view. Their instructors and students increasingly leaned towards holograms that could be enjoyed by their friends and families and—even more appealingly—sold to customers.

Just as importantly, holograms, not lasers, became the central attraction of these more coherent public shows.[9] During the mid-1970s, such exhibitions sprouted around the world and expanded in scale. Major exhibitions of the period transformed holograms from largely a North American experience to an international phenomenon. They introduced holograms to audiences ranging from hundreds of visitors to tens of thousands. Important examples included *Holography '75* (New York, 1975), *Holografi: Det Tredimensionella Mediet* (Stockholm, 1976), *Whole Message* (Vancouver, 1976), *Light Fantastic* and *Light Fantastic 2* (London, 1977 and 1978), *Alice in the Light World* (Tokyo, 1978, Figure 9.2), *Holographie Dreidimensionale Bilder* (Berlin, 1979), *Olografia* (Rome, 1979) and *Hololight '79* (Groningen, 1979).

[6] Benyon, M. to SFJ, interview, 21 Jan 2003, Santa Clara, CA, SFJ collection.

[7] Death rays were a popular notion throughout the twentieth century and were repeatedly linked to the new technologies of high-voltage electricity by Nikola Tesla (1920s), to microwave radiation (and later radar) by Robert Watson-Watt (1930s), and eventually to research on directed beam weapons as part of Ronald Reagan's Strategic Defense Initiative (1980s). See, for example, 'The "death ray" rivals', *New York Times*, 29 May 1924, 1; Haley, C., 'Science fiction and the making of the laser', in: M. Higgins, G. Lightfoot, M. Parker and W. Smith (eds.), *Science Fiction and Organization* (New York: Routledge, 2001), pp 31–9.

[8] Finch College, Museum of Art, Contemporary Study Wing, T. McBurnett and E. H. Varian, 'N-Dimensional Space,' New York, 1970.

[9] A notable exception, however, was *Light Fantastic* (1977), billed as 'Lasers and holography at the Royal Academy of Arts'. As the more recognizable and culturally-embedded of the two technologies, lasers continued to play a role in infecting the public imagination for holograms.

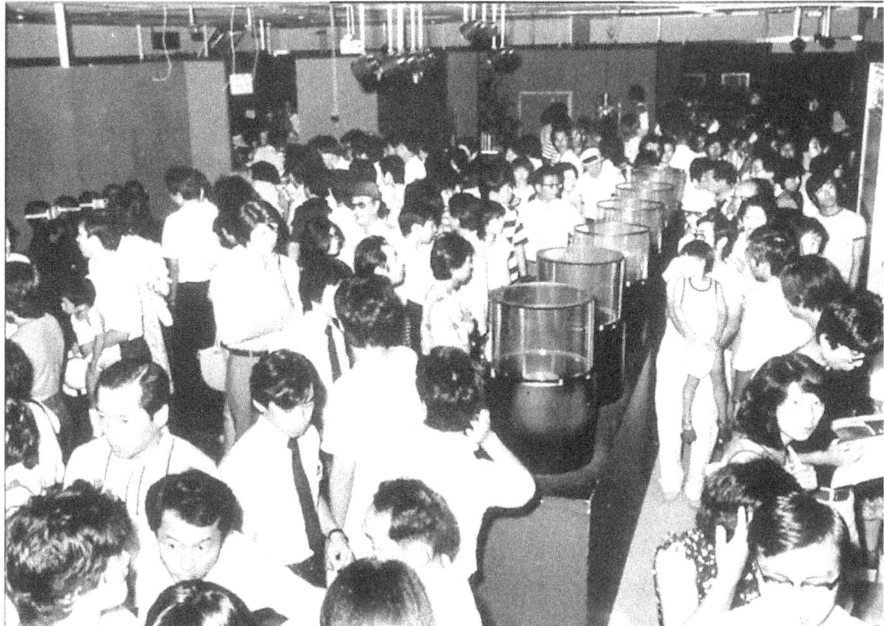

Fig. 9.2 *Alice in the Light World* exhibition, Tokyo, 1978. (Photo courtesy of I. Sakane).

British newspaper coverage peaked in 1977, the year of the *Light Fantastic* exhibition at the Royal Academy of Art in London. One newspaper reported that it was visited by some 3800 people per day, with four times as many more queuing in vain. Its effusive media reviewer adopted the same breathless optimism that had been invented a decade earlier in Ann Arbor:

> The next ten years could well become the Decade of Holography. Recent advances in three-dimensional photography created by means of lasers are such that a general invasion of film and theatre design, advertising, etc. can safely be predicted for the next couple of years. Wagner's Ring Cycle and Star Wars II and III—to mention only the grander projects—will thus contain new dimensions of illusion. Holographic television is only a few years away.[10]

Popular culture would be revitalized by holographic cinema, promised *Popular Science*.[11] And the following year at a Science Museum exhibition on lasers and holograms—attracting some 90,000 visitors—a London weekend columnist mused:

> There was a hologram of Professor Gabor, one of the originators, which was life-sized and so real that I began to grasp that the actual danger of the laser was not that it would burn us all up, but that it would usher in an era when illusion and reality could become almost

[10] Bowen, M., 'Laser days', *Guardian*, 12 Jan 1978, 8.

[11] Walton, S., 'Holography gets closer to your neighborhood theatre', *Popular Mechanics* 148 (1977): 94.

indistinguishable . . . Note to editor: can I please have a hologram at the top of my column at the earliest possible moment? I rather doubt whether an ordinary drawing will be compulsive enough by 1984.[12]

Even high culture was susceptible to such enthusiasms. A special issue of *Theatre Design and Technology* was devoted to state-of-the-art holography and laser projection which would be 'the theatre technology of the future'.[13] An art reviewer for *The Spectator* argued that as '[p]hotography led to moving pictures; holography will as inexorably lead to moving sculptures' and that 'the implications of the technique for films, television, theatre and, not least, the visual dreams of painters and sculptors are extraordinary . . . [and], potentially, the biggest visual art thing going'.[14]

The momentum of this impending future drove the major exhibitions of the following decade: 1982 *Space-Light* (Sydney, Australia, 1982); *Light Dimensions* (Bath, UK, 1983; London, UK, 1984); *Licht-Blicke* (Deutsches Filmmuseum, Frankfurt, Germany, 1984); *Images in Time and Space* (Montreal, 1987). As with circuses and vaudeville acts, fresh viewers were needed for spectacles that paled with familiarity. Many hologram shows were travelling exhibitions, garnering tens of thousands of visitors over several weeks at each venue. The format of the late 1970s exhibitions had set the standard. The hologram experience was to be an amalgam of art, technology, spectacle and mystery, with variety of subject and technique being key. This amalgam of attractions sustained a generation of exhibition-goers.

Cultural engagement with holograms took a different turn in the Soviet Union. There, reflection holograms dominated public displays. By the late 1960s, hologram researchers such as Yuri Denisyuk were being feted in Soviet newspapers and magazines.[15] During the 1970s research centres in Moscow, Leningrad and Kiev developed recording materials and techniques for uniquely bright and vivid holograms that remained the envy of western holographers.

Public exhibitions in Eastern Europe began in the mid-1970s but surpassed audience numbers in the western world. As news stories in *Pravda* and *Isvestia* reported, Ukrainian holograms were exhibited in the Crimea in 1976, with long queues of visitors swelling museum attendances. Installed in dedicated buses and trains, travelling exhibitions of holograms reached smaller population centres. An estimated one million visitors saw such displays every year in the Ukraine alone, and—as with American, British, French and Canadian shows—travelling exhibitions were exported abroad. Through the 1980s, Ukrainian holograms were shown in London, Yugoslavia, Italy and Korea.[16]

Yet holograms for the masses required constant promotion. Their dissociation from lasers made holograms less intimidating than they had been for earlier audiences but also

[12] Barker, D., 'Laser affair', *Guardian*, 4 Mar 1978, 15.

[13] *Theatre Design and Technology*, Fall 1979: special issue.

[14] McEwen, J., 'Art brought to light', *The Spectator*, 16 Apr 1982: 25.

[15] For example, Denisyuk, Y. N., 'Holography', *Soviet Union* 9 (1970): 12–4.

[16] Markov, V. B., 'Display and applied holography in culture development', in: *Holography: Commemorating the 90th Anniversary of the Birth of Dennis Gabor* (Bellingham, WA: SPIE, 1990), pp 268–304.

less engaging. Just as Margaret Benyon had struggled to show holograms to her audiences in 1969, a travelling exhibit designed by the New York Museum of Holography a decade later devoted a page to instructing audiences:

> Holography is changing the course of visual literacy because it forces us to re-evaluate our visual traditions in order to use it effectively. For instance, you don't look at a hologram—you look into it. A holographic image can project in front of the plate (a 'real image'), in back of the plate (a 'virtual image'), or it can straddle the plate (an 'image plane'). Several images can exist in any number of combinations of positions relative to the film plate . . .
>
> The holograms in this exhibition have been installed at the national average eye level. The optimum viewing occurs when your eye level is at the center of the film or plate, and when you are standing about an arm's length away from the work. If you are taller, bend down a little or step back a little farther. If you are shorter, come forward more. But remember, one of the fascinating things about holograms is that they look different from every angle and distance. So experiment. By approaching the work at different angles and heights, you will see different views, changing perspectives and spatial relationships, and sometimes shifting colors. This is all part of the excitement of the viewing experience. In order to enjoy and fully experience the dimensionality, we encourage you to bob up and down and move from side to side. Look from a distance. Then examine closely. And by all means, reach out and touch them. This is a new experience for the eye and mind, and they need reinforcement if they're going to believe what's in front of them.[17]

Between the lines, the exhibitors' laboured explanations were intended to forestall viewer complaints. Not only was the imagery unexpected, it could be hard to see. Unlike a window, these white-light holograms often had a restricted field of view. A parent might be unaware that her children could not see an image hung at adult height at all; bemused older people, unable or unwilling to bend enough, might pass a hologram display intended for children without a further glance. Even in a public space, this public spectacle was too often unshared or misunderstood. The experience, it seemed, was more akin to peering through a coin-operated telescope at a scenic location than it was to an awe-inspiring experience.

9.2 HOLOGRAMS AS ART

Harriet Casdin-Silver's atelier was on the fifth floor of a converted warehouse in the Fort Point district of Boston. Harriet, 78, was with her husband and an assistant who was locating holograms in the neatly arranged storage when I arrived. She spoke about her holograms and her place in the field, deftly enquiring about the current activities of other artists that I had visited. 'I love laser pieces', she told me; 'nothing is as mysterious feeling—when you get to the white light and all the rainbows, you don't get the same feeling'. Inhabiting a comfortable niche between science and art, Casdin-Silver's greatest impact had been in linking those cultures through her exploration of new types and uses for holograms.

Boston, 2003

[17] Jackson, R., *Through the Looking Glass: An Exhibition of Three-Dimensional, Floating Images Made With Lasers* (Oregon Museum of Science and Industry, 1980), quotations pp 4 and 6.

The awkward relationship between hologram exhibitions and lasers was paralleled by the diffident relationships between producers and their varied products. Hologram exhibitions brought together innovative artists, ambitious artisans and technologists. Audiences, too, were heterogeneous: art critics, technical enthusiasts, followers of the avant-garde and consumers vied to assess the meaning and value of holograms.

Each had distinct motivations. Technologists sought continual improvement in their holograms to extend their enthusiasms to larger audiences. Artists were eager to explore the new medium, sometimes forming uneasy alliances with university scientists or corporate labs to learn the requisite art. And—much like the stereoscope firms a century earlier—artisans connected these poles to create commercial imagery that could appeal through novelty and aesthetics. The wider public had been conditioned by the visual culture of the previous century: bemused by optical magic, captivated by innovation and excited by the latent potential of new technologies. And nearly all producers and consumers of holograms had been raised during the postwar period when modernity so effectively seduced a generation with expectations of accelerating technological advance.

Like the consumption of holograms themselves, holographic art was cosmopolitan and grew from prior cultural roots. The Machine Age had made technologies aesthetically pleasing as well as functional. The commercial design style known as 'moderne' between the wars (and later dubbed 'art deco' from the late 1960s) had amalgamated new materials, forms and colours into starkly geometrical patterns or streamlined shapes. Everyday items from vacuum cleaners to radios became *objets d'art* that expressed speed, power and modernity.

Such ubiquitous art could also portray the sophistication of technologies by hinting at their human dimensions. From the 1940s, a handful of commercial artists such as Boris Artzybasheff had created advertisements that anthropomorphized commercial machine products for companies such as Shell Oil, Avco Manufacturing and Xerox. His images turned lathes, helicopters and early computers into unsettling life forms—a visual forecast of the postwar field dubbed cybernetics.[18] Such depictions, popularized on the covers of current events magazines such as *Time*, added to the growing connections between art and technologies after the Second World War, as suggested by the depiction of new inventions through graphic arts. And, as with amateur enthusiasms, *Scientific American* magazine became a showpiece for such advertising art, particularly when it expressed human aspirations, dreams and power via new technologies.

These links between new technology and popular culture were extended to fine art by collaborations between engineers and artists during the 1960s. Among the more influential were two artists, Robert Rauschenberg (1925–2008) and Robert Whitman (b. 1935), and two engineers, Billy Klüver (1927–2004) and Fred Waldhauer (1927–93), who organized performance art events that brought together new technologies (e.g. fog machines

[18] An early exposition of the field was Wiener, N., *Cybernetics and Society: The Human Use of Human Beings* (Boston: Houghton Mifflin, 1954). For historical analyses of early cybernetics, see Heims, S. J., *Constructing a Social Science for Postwar America: The Cybernetics Group, 1946–1953* (Cambridge, MA: MIT Press, 1993) and Gerovitch, S., *From Newspeak to Cyberspeak: A History of Soviet Cybernetics* (Cambridge, MA: MIT Press, 2002).

and video projection), new materials (e.g. Mylar) and sensory surprise (e.g. electronically-generated noise as music and disorienting optical effects). They founded 'Experiments in Art and Technology' (E.A.T.) in 1967 as an organization to extend these artistic ideas.[19] Supplied by technologists, these new media provided a new palette for art.[20]

Such experiments influenced the first generation of holographic artists. Boston artist Harriet Casdin-Silver (1925–2008), in particular, started by employing lasers and later holograms in similar dynamic art installations:

> This is how I got into holography. I was looking for more sophisticated lighting. I used all sorts of lighting—strobe lighting, performance and sound—it was a performance . . . So I was looking for more sophisticated lighting for a stainless steel installation piece called Exhausts, and I went off to American Optical Company because I knew they were doing holography there, but I wanted a laser; I wanted to borrow a laser in my stainless steel. They invited me in to learn this new crazy thing they had already done—they were doing 4 × 5 inch holograms of rabbits—and I was really intrigued, and thought it would give me more to integrate with my other stuff.[21]

The same year, the Center for Advanced Visual Studies (CAVS) was founded by artist György Kepes at the Massachusetts Institute of Technology (MIT) to encourage collaborations between art, architecture and urban planning. Under its second Director, Otto Peine, the Center's remit of technological art made CAVS a focus for hologram projects, sponsoring Casdin-Silver, Dieter Jung (b. 1941), Betsy Connors (b. 1950), Paula Dawson (b. 1954) and other artists during the 1970s and 80s. The Center was eventually directed by Stephen Benton (1941–2003), a professor of physics at MIT who was a major contributor to the field of display holography from the 1970s. Most of the first generation of holographic artists—and the largest numbers practising to date—were born in the postwar baby boom.[22]

Oral histories suggest that these bodies and cultural currents inspired a handful of other artists to become interested in holograms as a medium for art. Lloyd Cross and sculptor Jerry Pethick, like Casdin-Silver, adopted Rauschenberg's 'found-art' aesthetic and E.A.T.'s mixed media, combining holograms with sound and laser displays.[23] Their first exhibitions of 1969 highlighted the laser light shows, but holograms were the centrepiece of their first exhibition in New York City, *N-Dimensional Space*, which then toured museums in Rochester, Syracuse and Albany, New York through 1970. 'At that point', reflected Cross, 'we were definitely into holography'.[24]

[19] Klüver, B., J. Martin and B. Rose (eds.), *Pavilion: Experiments in Art and Technology* (New York: E. P. Dutton, 1972).

[20] Hayward, P., *Culture, Technology and Creativity in the Late Twentieth Century* (London: J. Libbey, 1990); Popper, F., 'The place of high-technology art in the contemporary art scene', *Leonardo* 26 (1993): 65–9.

[21] Casdin-Silver, H. to SFJ, interview, 3 Jul 2003, Boston, MA, SFJ collection.

[22] A significant number of the first holographic artists—Rudie Berkhout, Brigitte Burgmer, Marie-Andrée Cossette, Setsuko Ishii, John Kaufman, Sam Moree, Al Razutis, Dan Schweitzer, and Fred Unterseher—were born in 1946, the very year that Dennis Gabor conceived holography.

[23] Gorglione, N., 'Forms of light: a personal history in holography', *Leonardo* 25 (1992): 473–80; Whittaker, J., 'On the beam . . . in color!', *Detroit Free Press*, 10 May 1970, 4-D.

[24] Cross, L., 'The Story of Multiplex', transcription from audio recording, Naeve collection, Spring 1976.

A handful of artists—particularly sculptors and performance artists—began to experiment with holograms. Bruce Nauman (b. 1941), who had studied mathematics and physics at university, explored combinations of sculpture, cinema, video and performance in his artistic works and, in 1968, used the pulsed-laser facility at Conductron Corporation in Ann Arbor to produce a series of holograms called *Making Faces*. His holograms captured a frozen performance:

> The idea of making faces had to do with thinking about the body as something you can manipulate. I had done some performance pieces—rigorous works dealing with standing, leaning, bending—and as they were performed, some of them seemed to carry a large emotional impact.[25]

The following year he returned to record a series of holograms with more intense laser pulses, *Second Hologram Series: Full Figure Poses (A-J)*. The grimacing self-portraits were displayed as three-dimensional holograms in a handful of venues (including the *N-Dimensional Space* exhibition) but in 1970 Nauman reused photographic images of the first series to generate conventional screenprints. These episodes were his only experiments with holograms.[26]

Salvador Dali, too, dabbled in holograms. He had flirted with science throughout his career, often vaunting the jargon and mystical dimensions of contemporary scientific advances. As he told one acquaintance, 'I swim between two varieties of water: the cold water of art and the warm water of science. I am the invisible line that separates the black from the white on a chessboard'.[27] During the early 1970s, Dali conceived scenes for sculptural models and, with collaborator Selwyn Lissack, had them produced at Conductron Corporation as his first experimental holograms. In 1973, he was the first client for a cylindrical animated hologram made by Lloyd Cross's Multiplex Company.[28] The hologram, *First Cylindric Chromo-Hologram Portrait of Alice Cooper's Brain*, was a complex creation depicting the rock musician Alice Cooper adorned with jewellery and a hand-held sculpture. The piece expressed the shock typical of Dali's painted subjects and Alice Cooper's performances, much as the earlier visual arts of stereograms and cinema had done before them. Dali's self-promotions were performances, too, in which his holograms were contributing players.

For other artists, the optical properties of holograms were a greater attraction. Margaret Benyon, a British artist who was exploring depth illusion in her paintings, took up holography at about the same time. Having heard of holography through a newspaper story, she was intrigued by the connections with her own art. Unlike Nauman,

[25] Cordes, C., 'Bruce Nauman: Prints 1970–89', New York, 1989.

[26] As with all holographic art, photographs of Nauman's holographic images, particularly the first series, are far more numerous than exhibitions. See, for example, Nauman, B., *Artforum*, 9 (4), Dec 1970: cover art.

[27] Thurlow, C., *Sex, Surrealism, Dalí and Me: The Memoirs of Carlos Lozano* (Cornwall: Razor Books, 2000), p. 74.

[28] Lissack, L. and S. Lissack, *Dali in Holographic Space* (e-book: CreateSpace, 2012); Cross, 'Holostories', op. cit.; Moore, Lon to J. Ross, interview, 1980, Los Angeles, Ross collection; Unterseher, Fred to SFJ, interview, 23 Jan 2003, Santa Clara, CA, SFJ collection.

she learned how to make holograms herself by spending time in physics labs, becoming probably the first independent art holographer and a prominent contributor to the art.[29]

For yet others, small-scale modelling—sculptures smaller than a breadbox—proved ideal for the holograms of the period. After encountering Billy Klüver and seeing a laser beam for the first time at Bell Laboratories, the Swedish painter/sculptor Carl Fredrik Reuterswärd (b. 1934) began exploring holography in 1969 to create a considerable body of work.[30]

Artists transformed holograms technically. Harriet Casdin-Silver convinced Stephen Benton, its inventor, that the rainbow hologram was a useful variant for aesthetic reasons. Other artists through the 1970s experimented with combining multiple exposures of the rainbow process to generate 'pseudocolour' holograms that would reveal multiple colours when lit by a white light source.

Over the following three decades, a number of artists having reputations in other media experimented more briefly with holograms.[31] A few of the early adopters, such as Casdin-Silver and Benyon, continued to focus on fine art holography over their professional careers. An even smaller number of dealers and galleries were favourable to exhibiting holographic art, notably the Leo Castelli Gallery in New York, which exhibited Nauman's work in a 1969 show, *Bruce Nauman Holograms, Videotapes and Other Works*, and the Knoedler Gallery in New York, which displayed Dali's holograms in a 1975 exhibition, *The Threshold of the Optical Renaissance*. The pieces have seldom been exhibited since, except in museums dedicated to Dali's body of artistic work in Florida and Spain.

High art, artisanal products and do-it-yourself holograms coexisted in the venues that would show them. In 1974, the *Village Voice* could report:

> America's first holographic museum has opened at 120 West 20th Street in conjunction with the New York School of Holography . . . There is even a small vibration isolation table at the second floor gallery complete with its own laser, so you can fool around with your own holograms. Also on display are the works of old masters of this new art, including a couple of three-dees by Salvador Dali.[32]

This uncomfortable jostling of fine art with amateur creations, technological promotion and commercial products helps explain the tensions surrounding holograms as an art medium. Heterogeneous ambitions and audiences complicated the popularity of art holograms.

Fine artists themselves resented interlopers from technical fields, arguing that they diluted quality and blurred boundaries. Casdin-Silver recalled of a 1975 hologram exhibition,

> I wanted to take my art out of that show . . . I thought it stank to high heaven. There was so little real art in it . . . Typical people who had six weeks of holography called themselves artists.[33]

[29] Benyon, M. to SFJ, interview, 21 Jan 2003, Santa Clara, CA, SFJ collection.

[30] Bjelkhagen, H. I. to S. Johnston, interview, 18–9 Sep 2002, SFJ collection.

[31] These included artist/sculptor Louise Bourgeois (1911–2010, during the late 1990s), sculptor/performance artist Eric Orr (1939–98, in 1995), Ed Ruscha (b. 1937, from 1998), James Turrell (b. 1943), choreographer/musician Simone Forti (b. 1935), Michael Snow (b. 1939) and filmmaker Stan Vanderbeek (1927–84).

[32] Smith, H. and B. Van der Horst, 'Scenes', *Village Voice*, 14 Feb 1974: 2.

[33] Casdin-Silver, H. to SFJ, interview, 3 Jul 2003, Boston, MA, SFJ collection.

Art critics, too, were an early hurdle. A brief mention in print of the first London show by artist Margaret Benyon in 1970 was typically diffident. 'The stereoscopic effect is perfect. You cannot believe the objects are not really there . . . the results are startling and sometimes beautiful . . . [but] from the art point of view it is still in the pilot stage'.[34]

The *New York Times* was a conduit of metropolitan disdain for art holograms. The reviewer of one of the first American shows at an art venue summarized in 1970,

> I wish I could wax more enthusiastic, as they say, about the esthetic potential of holography which, as every preschooler seems to know, is the making of three-dimensional photographs by laser beam. But while 'N Dimensional Space', a show of holograms by artists at the Finch College Museum of Art/Contemporary Wing, is earnest and enlightening, it doesn't go far toward convincing me.[35]

Half a decade later, arts reviewer Hilton Kramer (1928–2012), chief art critic of the *New York Times* and *The Nation*, underlined the judgement: 'There is always something disconcerting in the spectacle of immensely sophisticated technology—which artists sometimes call "science"—serving as the vehicle for some perfectly trivial conception'.[36] A political conservative, Kramer was a staunch defender of high culture and an opponent of what he perceived as a left-wing bias in both art and art criticism. His dismissal, broadly echoed by the American art establishment, was influential in restricting hologram exhibitions to minor venues, provincial contexts and public entertainments.

Subsequent critics took a postmodern turn, increasingly challenging even the fascination with science that had been at the core of modernist praise for holography. A *Chicago Tribune* critic opined that the technology,

> Does not yield many holographic possibilities that would effectively transcend that 'miracle of science' status [. . .] Until it is firmly established, holography will continue to perform its one spatial variation on some of the most traditional art forms of all. That is hardly enough to justify the revolutionary rhetoric.[37]

Thus the technical maturity, aesthetic insights and modernist subtext of art holograms were successively criticized by the art world. But these experiences were not merely an American phenomenon. The production of holographic art—unlike the first wave of technical experimentation and publicity about holograms—was international, as evidenced by significant contributors such as Pethick (Canada), Benyon (Britain), Reuterswärd (Sweden), Rudie Berkhout (1946–2008, Netherlands), Paula Dawson (Australia), Dieter Jung, (Germany), Setsuko Ishii (b. 1946, Japan) and Eduardo Kac (b. 1962,

[34] Gosling, N., 'Beaming pioneer', *Observer*, 1 Mar 1970, 33.

[35] Glueck, G., 'Works made with aid of laser on display at Finch: art holograms in their infancy', *New York Times*, 25 Apr 1970, 24.

[36] Kramer, H., 'Holography: a technical stunt', *New York Times*, 20 Jul 1975, D1–D2. See also a similar review, Hess, T. B., 'Hocus-pocus and hocus-focus', *New York*, 8 (33), 18 Aug 1975: 70–1.

[37] Artner, A. G., 'Rhetoric, not results, at holography show', *Chicago Tribune*, 12 Jun 1977, 37.

Brazil).[38] While initial public interest was high, the longer-lasting critical disparagement was also similar in each country. In 1978, musician and critic George Melly gamely provided a segment on 'A Beginner's Guide to Holograms' for the arts programme, *Arena*, on Britain's BBC2 television. As an introduction to the *Light Fantastic 2* exhibition at the Royal Academy of Arts, the late-night programme aimed to boost attendances, but appears to have garnered little audience reaction. And a year later, another British reviewer dismissed the most widely admired holographic artworks of the day at a Liverpool exhibition:

> So far holo-artworks are little more than test cards. Note, however, the image of somebody in a foetal position, her buttocks looming in wondrous multicolour, secured by Harriet Casdin-Silver, Fellow of the Advanced Visual Center [sic] at MIT. Clearly there's a great future for holography in the field of adult home entertainment.[39]

Another reviewer of the show noted:

> The images themselves are rather a job lot. The Guinness bottle, the inevitable models from Star Wars, the pseudoscopic skull, the model theatre set are not in themselves very interesting objects . . . But the hologram merely reproduces an image. Once the novelty has worn off, it is only as interesting as the image it reproduces. To the non-technologist it's merely another sideshow . . . A truly new visual vocabulary, be it technical like the cinema, or conceptual, like the Cubist movement, enlarges the capacity to order and makes sense of our experience. Holography does not.[40]

British art critic Edward Lucie-Smith later argued that, while art holography had attracted interest during the heady 1960s, it fallen out of step with the major trends in contemporary art. As holographic artists became more numerous in the early 1970s,

> Op Art was just going out of fashion—and anything 'optical' was suspect . . . Holographers who aspired to be artists boasted of the fact when they managed to produce six or seven small images a year. This was entirely contrary to one important trend in the art of the 1980s and 1990s, which valued scale (vast environmental works were and are the height of fashion), untrammelled spontaneity, and the use of so-called 'poor' materials.[41]

[38] Eastern Europe, by contrast, produced little holographic art before the fall of the Soviet Union. Instead, holograms were employed as surrogates for cultural treasures, as discussed in Section 9.3.

[39] Feaver, W., 'Crafty stuff: art', *Observer*, 16 Dec 1979, 18. Even adult entertainment proved to have limited appeal. The Multiplex Company in San Francisco produced a series of erotic short subjects (typically 15 s long), in their ubiquitous cylindrical hologram format during the mid-1970s, representing some 20% of their stock catalogue (Section 8.3). More explicit Multiplex imagery was trialled by Peter Claudius during the late 1970s, but found brief application mainly as a novelty pay-for-display viewing device in San Francisco adult theatres. During the 1980s, when pulsed lasers became more available and public interest was high, many commercial holographers dabbled in holographic erotica (Munday, R. to SFJ, interview, 31 Mar 2004, Richmond, UK, SFJ collection; 'Holographic playmates', *Holography News*, 15 (2), Mar 2001: 4). During the peak of hologram exhibitions and public recognition, men's magazines tried to boost circulation with photo-spreads of two-colour (anaglyphic) stereo images; see, for example, Paul Raymond Publications, '3D Hologram Issue: Amazing hologram 3D centre set with free sex spex', *Men Only*, 48 (4–6), Apr–Jun 1983: cover.

[40] Austin, D., 'Exhibition: Light Fantastic 2', *New Scientist*, 19 Jan 1978: 171.

[41] Lucie-Smith, E., 'A 3D triumph for the future', *The Spectator* 277 (1996): 57–8.

The major museums of modern art (e.g. Tate Modern in London, Peggy Guggenheim in Venice, Moderna Museet in Stockholm) did not collect holograms.[42] Through the 1990s and 2000s, critical rejection of art holograms was much lamented by artists and periodically rehearsed by the critics themselves, but the positions of the two subcultures and their arguments were enduring.[43]

9.3 HOLOGRAMS AND HERITAGE

Artisans, engineers and scientists, it appeared, did not make good artists. Diluted of both aesthetically interesting and freshly spectacular subjects, exhibitions failed to capture the sustained attention of subsequent reviewers, who more frequently ignored holographic art shows than criticized them. But while art holography and exhibitions of holograms arguably attracted their greatest attention during the 1980s and 1990s, hologram displays in museums had a slower and continuous rise in popularity (Figure 9.3).

The appeal of holograms in a museum context arose from the temporary exhibitions. When public confidence in progress was at its highest point, hologram displays could illustrate rapid improvement and a confident future. Public exhibitions consequently emphasized not merely visual surprise and artistic aesthetics, but also a history of the subject that depicted progressive technological perfection. Similar themes had long been popular for displays at the National Museum of American History and National Air and Space Museum (Washington DC), the Science Museum (London) and the Deutsches Museum (Munich), where accounts of engineering innovation could be allied with master-narratives of scientific progress and national identity.[44]

The first fixed exhibitions of holograms adopted a similar modernist theme by identifying holograms as the next generation of photographic processes.[45] The New York Museum of Holography (1976–92), which for a time positioned itself as *de facto* ambassador for the subject, incorporated an historical section in its travelling exhibits and its permanent attractions. The theme was developed by seeking donations of 'historic'

[42] Some smaller and specialist museums have included holographic art in their permanent collections, including the Karl Oesthus Museum (Hagen, Germany) and the Butler Institute of American Art (Youngstown, OH) (Lauk, M., 'Holography in museums—a modern Sleeping Beauty syndrome', presented at *Holography in the Modern Museum*, Leicester, UK, 2008).

[43] See, for example, Lightfoot, D. T., 'Contemporary art-world bias in regard to display holography: New York City', *Leonardo* 22 (1989): 419–24; Hagen, C., 'The case for holograms: the defense resumes', *New York Times*, 29 Nov 1991, C24, Barilleaux, R. P., 'Holography and the Art World', *Leonardo* 25 (1992): 417–8; Pepper, A., 'When did creative holography die?', *Journal of Holography and Speckle* 3 (2006): 80–3.

[44] A postwar article, for example, described museums as 'a powerful means of publicizing scientific achievements', and illustrated exhibits on penicillin, low temperature research, nuclear energy and the development of polythene as examples ('Museums and scientific progress', *Discovery*, 8 (8), Aug 1947: 228–9).

[45] Johnston, S. F., 'Attributing scientific and technical progress: the case of holography', *History and Technology* 21 (2005): 367–92.

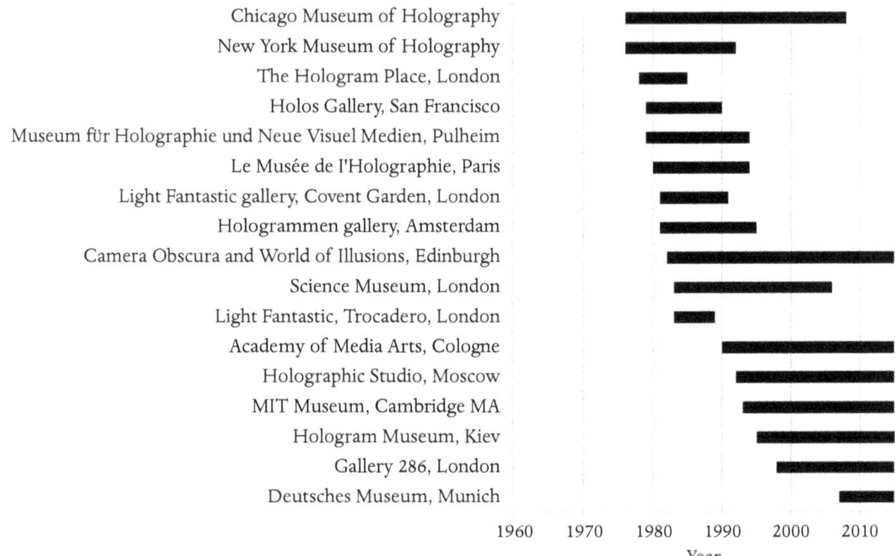

Fig. 9.3 Galleries and long-term exhibits of holograms.

holograms from their producers.[46] In Britain, *Light Dimensions: The Exhibition of the Evolution of Holography*, held at the National Photographic Centre of the Royal Photographic Society (RPS) in Bath, UK, in 1983 again revived the putative links between early photography and current holography, confidently predicting a similar technical, social and cultural advance. The RPS, in fact, consolidated its stake in the field by launching a Holography Group that year.

By the mid-1980s, however, the exhibition boom was faltering. Some of the exhibition collections had a second life as permanent exhibits in museums, such as London's Science Museum (Figure 9.4). Its showpiece display was an exciting addition to the Optics Gallery, which had had its last significant update following the war and was crowded with traditional devices such as lighthouse lenses and industrial instruments. The holograms were a selection from the third successful British holography exhibition, *Light Dimensions*. After attracting impressive audiences in Bath in 1983, it had been moved for a month-long display at the Science Museum before being scaled down for installation upstairs. Over the following 23 years, the hologram displays suffered chemical deterioration (typical of the experimental processes that had yielded bright images), periodically failing light sources and steadily declining audiences. The hologram exhibit, along with the entire Optics Gallery, was removed in 2006 and relegated to the Museum stores.

Commercial museums increasingly competed for audiences. Besides New York's Museum of Holography, they included the Fine Arts Research and Holographic Center in Chicago (founded 1978), the Museum für Holographie und neue visuelle Medien

[46] Lancaster, I. M., 'The Museum of Holography – its role and policies in a changing environment', presented at *Practical Holography II*, Los Angeles, CA, 1987.

Fig. 9.4 Entrance to the Optics Gallery, Science Museum, London, 1988, displaying a pulsed laser hologram of Dennis Gabor, inventor of the principle of holography. The hologram, a Leith–Upatnieks (laser transmission) hologram, silver halide on glass, 18" × 24", was recorded at Conductron in 1971 (Science & Society Picture Library, Science Museum, London).

(Pulheim, Germany, 1979) and the Musée de l'Holographie (Paris, 1980). From the travelling exhibitions, they inherited not just the holograms, but the narrative of progress, too. The hologram was thereby recast as an historical object.[47]

Hologram 'museums' were also modelled on art museums. Excluded by the art world but paralleling it, they attracted paying customers with compelling imagery.

[47] See also Johnston, S. F., 'Why display? Representing holography in museum collections', in: P. Morris and K. Staubermann (eds.), *Illuminating Instruments* (London: Smithsonian Museum Scholarly Press, 2010), pp 97–116.

Examples included the Light Fantastic Gallery (Covent Garden, London, 1981 and Trocadero, London, 1983) and The World of Illusions (Edinburgh, 1983). Such commercial galleries struggled, though, to supplement their incomes by sales of art holograms.[48] For a more recent example of an exhibition dedicated specifically to art holography, see Figure 9.5.

On the whole, traditional museums, apart from those vaunting national technologies, shunned holograms.[49] But holograms in museums also held promise as substitutes for precious objects. The use of museum proxies has a long history. London's Victoria and Albert Museum, opened in 1873, still has a gallery devoted to plaster-cast replicas of architecture, sculptures and even Trajan's column. Regional museums such as the

Fig. 9.5 Abstract multicolour reflection holograms by Iñaki Beguiristain, *Pure Holography* exhibition, Brighton, 2014 (Holograms from left to right are *Light Weave*, 8×10, 2014; *Illuminating the Invisible*, 8×10, 1999; *Colour Wave*, 30×40, 2001; *Nucleus*, 8×10, 2014; *Coloured Lines*, 8×10, 2002; *3D Shadows*, 8×10, 2014. All are H1 / H2 pseudocolour reflection holograms, silver halide on glass. Photo by John D. Brown, courtesy of I. Beguiristain).

[48] 'Number of galleries showing holograms continues to grow', *Holosphere*, 6 (1), Nov 1977:; Zellerbach, G. A., 'Selling holographic art: an analysis of past, present and future markets', *Proceedings of the SPIE* 1600 (1991): 446–54; Zellerbach, G. A., 'Comments on the Market for Holographic Art', *Leonardo* 25 (1992): 419–20. A typical commercial gallery exhibit priced its art holograms between $750 and $8800 in 1999 (The Green Art Gallery & New England Sculptor's Guild, 'Timeless Visions: Holograms from the Collection A D 2000', Guildford, CT, 1999, G. Zellerbach collection).

[49] The Victoria and Albert Museum, Britain's paramount museum of culture, established a hologram collection policy in the 1980s, but this did not outlive the tenure of its curator. Even the putative historical link with photography was not universally promoted: the National Museum for Photography, Film and Television (later the National Media Museum, in Bradford UK) did not collect or display holograms.

University of Illinois at Urbana-Champaign often dedicate significant fractions of their exhibition space to such reproductions, but stand-ins are common even at home; the Museum of Plaster Casts at the University of Aristotle (Thessaloniki) displays copies of ancient Greek objects.

In the Soviet Union, where art holography had not gained a foothold, this museum usage became a major application: national treasures would be democratized with holograms serving as their surrogates (Figure 9.6).

To be used as proxies for the museum objects, hologram imagery had to be as realistic as possible. Denisyuk reflection holograms were ideal for the purpose. They could be recorded by a simple optical set-up, viewed with an inexpensive light source, and yielded a faithful reproduction in three dimensions. Russian researchers had developed high-quality photographic emulsions to optimize the image brightness and clarity. When their holograms were displayed at Strasbourg, France, in 1976, and Ulyanovsk in 1979, Western observers were chastened to confirm their superior technical quality. Stephen Benton

Fig. 9.6 Reflection hologram of an 18thC silver gilt icon of the Kazansky virgin from the Kiev-Pechersky Lava collection (*Holography in the USSR: the Art and Science of the Soviet Union* exhibition catalogue (1981), courtesy of V. Markov).

of Polaroid noted, 'it's one of the few times in my life when I've seen a hologram and it seemed like a glass case full of objects themselves'.[50] 'Real 3-D', while a marketing failure at Conductron, had found a home in Soviet museums.

The Ministry of Culture in the Ukraine supported a holographic laboratory under the direction of physicist Vladimir Markov from 1979, with holograms of heritage treasures displayed in permanent and travelling exhibitions across the Ukraine.

The idea of employing holograms as museum surrogates had circulated in the west, but found only weak institutional support.[51] Despite the Ukrainian and Russian successes and refinements in holography that allowed strikingly realistic colour reproductions, western museums did not pursue large-scale programmes.[52] It could be argued that museum holograms cannot convey the impact of the rare objects themselves, or even that video or online images provide an adequate simulacrum for mildly curious museum audiences. The issues were likely political as much as technical or cultural. It may be that such facsimiles threatened to make museum collections and sites less unique, and challenged the autonomy of the institutions themselves.

9.4 HOLOGRAMS AND THE POSTMODERN

While exhibitions and museums told compelling tales of modernity, cultural critics generally considered holograms in postmodern terms. Where the public shows vaunted the progressive rationalization of society through technology, scholars began to assess how holograms reflected this culture. As Jean Baudrillard mused in 1981,

> Which type of objects or forms will be 'hologenic' remains to be discovered since the hologram is no more destined to produce three-dimensional cinema than cinema was destined to reproduce theatre, or photography was to take up the contents of painting.[53]

There were few critical audiences for holograms in 1970. Lloyd Cross and Jerry Pethick had sprinkled their media encounters with 'nonlinear' imaginings, trying to envisage how holograms might inspire individualistic perspectives (Section 8.2). Art critics had been the first to weigh in against them, often judging display holograms to fit poorly with contemporary art movements (Section 9.2).

[50] 'Soviet holography conference shows high amount of research activity', *Holosphere*, 7 (9), Sep 1978: 1–2.

[51] Bush, E. A., 'Hologram cylinders, 4 meters long, display rare objects without risk', *Holosphere*, 7 (10), Oct 1978: 1–2; Wuerker, R., 'Holography of art objects', *Proceedings of SPIE* 215 (1980): 167–71.

[52] Markov, V. B., 'Holography in museums – why not go 3-D', *Museum* 44 (1992): 83–6; Bjelkhagen, H. I. and D. Vukicevic, 'Color holography: a new technique for reproduction of paintings', *Proceedings of SPIE* 4659 (2002): 83–90; Bjelkhagen, H. I. and P. G. Crosby, 'Color holography and its use in display and archiving applications', *Archiving 2008 Final Program and Proceedings* 2008 (2008): 239–45; Markov, V. B., 'Holography in museums', *The Imaging Science Journal* 59 (2011): 66–74.

[53] Baudrillard, J. and S. F. Glaser (transl.), *Simulacra and Simulation* (Ann Arbor: University of Michigan Press, 1981), p. 104.

Postmodern evaluations over the following decade tended to be more receptive to holograms. Emerging in postwar France, postmodernism developed as a critical movement in continental Europe during the 1960s and beyond. Its analytical perspectives successively influenced literary criticism, the social sciences and philosophy by the 1980s. The public debut of the hologram was consequently well-timed; as a revolutionary visual medium, it provoked a new generation of critical evaluation just as the photograph, stereoscope, cinema and television had for earlier generations.

For postmodernists, confidence in the predictability of technological change appeared naïve. Science, they argued, could seldom be reduced to the accumulation of facts or divided into distinct investigations. Human powers and activities had more complex underpinnings. In place of scientific faith ('scientism'), postmodernists opted for analysing the cultural currents that traced more abstract forces.

But attempts to classify, compartmentalize and judge movements in art also attracted their criticism. The postmodernist perspective recognized that creative works could have multiple interpretations, and valued diversity of expression. Artworks—and even inventions—did not have fixed meanings. The motivations of their creators were implicit or reinterpretable, and viewers could discover their own meanings. Superficial meanings and order had to be *deconstructed* to liberate such creative works.

This challenging of modernist notions gave holograms a breathing space in culture. They did not have to exemplify corporate optimism or faith in the future. Instead, holograms might find varied, and even competing, uses. They would be as malleable as culture itself, co-evolving with it.

For postmodern scholars and less radical cultural commentators, holograms offered a fertile example of the contemporary world. As quickly as holograms had become visible to artists and wider publics, they came to represent or epitomize something about the spirit of the age. With varying degrees of success, these evaluations often challenged narrow beliefs in technical progress and its social consequences.

One of the first to notice holograms was Alvin Toffler (b. 1928) in his 1970 bestseller *Future Shock*. His book warned that technological society was driving a rate of change that would cause cultural earthquakes. His book joined a growing collection of analyses that forecast unanticipated changes, including *The Population Bomb, The Greening of America, Limits to Growth* and *Small is Beautiful*, each of which also became bestsellers.[54]

These studies introduced a more critical view of modernity than had been common after the Second World War. There had been precursors, certainly. Technology as an irresistible but largely positive social force became an explicit assumption for

[54] Erlich, P. R., *The Population Bomb* (New York: Ballantyne, 1968); Reich, C. A., *The Greening of America: How the Youth Revolution is Trying to Make America Livable* (London: Allan Lane, 1970); Club of Rome, *The Limits to Growth: A Report for the Club of Rome's Project on the Predicament of Mankind* (New York: Universe, 1972); Schumacher, E. F., *Small is Beautiful: Economics as if People Mattered* (New York: Harper & Row, 1973).

wartime and postwar scholars, politicians, administrators and the wider public.[55] Historian Jacques Ellul's *The Technological Society* had argued, though, that our drive towards inventive progress had run out of control during the twentieth century, leading to technologies themselves becoming autonomous and mutating in unpredictable and possibly undesirable ways—the notion of technological determinism. Political philosopher Herbert Marcuse's *One-Dimensional Man*, a favourite of the youth movement during the 1960s, contended that technologies and consumerism would be guided by governments and corporations to oppress and repress. And, from another disciplinary perspective, biologist Rachel Carson's *Silent Spring* documented the unexpected harms that DDT—the miracle pesticide of the Second World War and postwar world—had on ecosystems.[56]

What these diverse books shared was an increasingly critical vision of modern culture, identifying our faith in technologies and progress as a key failing. For the pessimists, the near future appeared overpoweringly dystopian; for the optimists, new technologies would be harnessed to create the future we desired. Amongst the latter was the inventor of holography himself, Dennis Gabor. In the early 1960s, while his holograms were judged an unsuccessful invention, his book, *Inventing the Future*, provided a more hopeful perspective.[57]

Alvin Toffler, too, suggested that technology could be controlled if we were attentive and prepared for change. He said rather little about holograms, but enough to characterize them as typical of forces responsible for a new fragmentation of society:

> As science expands and the scientific population grows, new specialties spring up, fostering more and still more diversity at this 'hidden' or informal level. In short, specialization breeds subcults . . .
>
> We can anticipate the formation of subcults built around space activity, holography, mind-control, deep-sea diving, submarining, computer gaming and the like. We can even see on the horizon the creation of certain anti-social leisure cults—tightly organized groups of people who will disrupt the workings of society not for material gain, but for the sheer sport of 'beating the system'.[58]

New technologies like holography were responsible, he suggested, for a new pluralism that threatened to destabilize the homogeneity of culture and society itself. For those

[55] Among the most influential were Bush, V., 'Science The Endless Frontier: A Report to the President by Vannevar Bush, Director of the Office of Scientific Research and Development', United States Printing Office, July 1945; Wilson, H., 'Conference address, Labour Party Annual Conference,' Scarborough, 1963, which oriented British government policy towards harnessing the 'white heat' of science and technology.

[56] Ellul, J., *The Technological Society* (New York: Knopf, 1964), first published in 1954 in France as *La Technique*. For an analysis of these cultural currents, see Winner, L., *Autonomous Technology: Technics-out-of-Control as a Theme in Political Thought* (Cambridge, MA: MIT Press, 1978). Marcuse, H., *One-Dimensional Man: Studies in the Ideology of Advanced Industrial Society* (Boston: Beacon Press, 1964); Carson, R., *Silent Spring* (Boston: Houghton-Mifflin, 1962).

[57] Gabor, D., *Inventing the Future* (London: Secker & Warburg, 1963).

[58] Toffler, A., *Future Shock* (London: Pan Books, 1971), pp 263–4. For his focus on preparing for change, see the first of a series of follow-on books aimed increasingly at business managers: Toffler, A., *The Third Wave* (New York: Morrow, 1980).

pursuing these new cultural directions, Toffler's analysis seemed superficial and threatening. Artist Margaret Benyon later reflected:

> When Alvin Toffler first prophesied in 1970 that holography would become a 'subcult' in his book Future Shock I felt very indignant and insulted at being dubbed a 'fun specialist'. To me, holography deserved a more worthy future. But despite the best efforts of a few of us, a subcult is what it became. Once this happened, it became difficult for holographic artists to be taken seriously by the art world.[59]

Even more divisively, Toffler's views inverted those of the counterculture about holograms. For proponents of the counterculture, holograms had been a medium for holism. For Toffler, holograms represented cultural splintering.

While Toffler identified himself as a futurist and, later, a business visionary, more academic scholars began to engage with holograms and culture. Jean Baudrillard (1929–2007) discussed holograms as the embodiment of a *simulacrum*, or an inadequate representation of reality. He argued that its images are 'mercilessly detailed' yet empty—literally intangible—versions of the real thing with properties that are unlike those of a mirror or optical illusion. Instead of being a copy or double of something real, he suggested, holograms confer new Godlike powers:

> We dream of passing through ourselves and of finding ourselves in the beyond: the day when your holographic double will be there in space, eventually moving and talking, you will have realized this miracle. Of course, it will no longer be a dream, so its charm will be lost.[60]

Baudrillard argued that holograms exemplify modernity. They satisfy the desire for the most faithful and contemporary representations of reality, on the one hand, but also invent a highly artificial reality of their own that is *hyper-real*. The imagery is simultaneously mesmerizing, desirable and false, reflecting modern culture itself.

Umberto Eco (b. 1932) adopted the term for the updated English translation of his 1973 essay *Il Costume di Casa* and addresses similar themes. Holography, he claims, 'could prosper only in America, a country obsessed with realism':

> You look into a magic box and a miniature train or horse appears; as you shift your gaze you can see those parts of the object that you were prevented from glimpsing by the laws of perspective. If the box is circular you can see the object from all sides. If the object was filmed, thanks to various devices, in motion, then it moves before your eyes, or else you move, and as you change position, you can see the girl wink or the fisherman drain the can of beer in his hand. It isn't cinema, but rather a kind of virtual object in three dimensions that exists even where you don't see it, and if you move you can see it there, too.

Simultaneously fascinated by its modern magic and bemused by its pop culture, Eco sketches the growing division between such representations and individually-experienced reality. The San Francisco holograms of the 1970s were part of a cultural context that

[59] Benyon, M. to S. F. Johnston, email, 2 Feb 2003, SFJ collection.

[60] Baudrillard, J. and S. F. Glaser (transl.), *Simulacra and Simulation* (Ann Arbor: University of Michigan Press, 1981), p. 103.

included the *Ripley's Believe It Or Not* and *Movieland Wax* museums. As suggested by the book's original English fore-title, *Faith in Fakes*, Eco underlines cases 'where the American imagination demands the real thing, and, to attain it, must fabricate the absolute fake'.[61]

Both Baudrillard and Eco are early observers and critics of the popular engagement with holograms. They link the hologram's 'maniacal chill' and 'exactness of detail' to incongruous inauthenticity and deception.[62] Indicative of modern culture, they hint, holograms are a symptom, not a solution.

9.5 COLLECTING HOLOGRAMS

Holograms promised to become a mass cultural experience. Andy Warhol speculated in the mid-1970s about holograms in the home:

> Holograms are going to be exciting, I think. You can really, finally, with holograms, pick your own atmosphere. They will be televising a party and you want to be there, and with holograms you will be there. You will be able to have this 3-D party in your house. You will be able to pretend you are there, and walk in with the people. You can even rent a party. You can have anybody famous that you want sitting right next to you.[63]

If exhibitions could generate a collective response, what about individual engagement? A 1974 in-flight magazine, reporting the growing activity in the field, mused:

> Imagine, for instance, the artsy/camp value of a hologram-generated art deco face peering up at its wearer from the toe of her shoe when illuminated by beams of concentrated white light in a boutique . . . Certainly it would be visually stimulating to wear a suit of defractory [sic] clothes while sitting on a defractory sofa surrounded by defractory walls and ceiling. There has also been talk of creating a hologram of a shapely nude body which one could wear on the garments covering one's own dumpy physique. However, one would then have to consider the consequences once out of the light source . . . holography has managed to keep the ennui-ridden hipsters fascinated, and there is no indication that the fascination is lessening. It is strange, maybe, to contemplate the bizarre juxtaposition of Dennis Gabor inventing the holograms as part of his quest to see the lone atom, and Zsa Zsa Gabor wearing it, but such are the times.[64]

From the mid-1970s, holograms became steadily more available via producers, retailers and artists for consumers who browsed technical catalogues or visited the expanding hologram exhibitions and museums. This first wave of products remained relatively exclusive, though: even the mass-produced holograms for hobbyists available from Edmund Scientific sold for at least $6 apiece, and required a $220 laser for appropriate lighting. Multiplex holograms sold for between $100 and $330, including display unit (prices that

[61] Eco, U. and W. Weaver, *Travels in Hyper-Reality: Essays* (London: Picador, 1986); quotations pp 3–4, 8.

[62] Eco, U. and W. Weaver, *Travels in Hyper-Reality: Essays* (London: Picador, 1986); quotation p. 10.

[63] Warhol, A., *The Philosophy of Andy Warhol (From A to B and Back Again)* (San Diego, CA: Harcourt, 1975), p. 158.

[64] Rolfe, L. and N. Lennon, 'Holography – new light on a new dimension', *Delta Airlines Sky*, Dec 1974: 23–9.

should be scaled by a factor of five for 2015 equivalents). To more demanding audiences, supplies of holograms were growing as entrepreneurs, trained in the first schools of holography, started to finance their enthusiasms through cottage industry and direct sales to a growing public. By contrast, few art holograms sold at all.[65]

Acquiring holograms became a new trend for niche audiences, but founded on familiar motivations. A small and exclusive audience collected holograms as art. As with more traditional forms of art, collectors were distinguished by their aesthetic sensibilities or expectations of rising value. While a handful of affluent individuals amassed collections, holographic art was never exorbitant in price or consensually judged. The majority of pieces either remained in the holographic artists' own collections or were exchanged between artists.

For an intermediate audience, amassing holograms had an appeal similar to that of creating *cabinets of curiosities* during the eighteenth century: the satisfaction of owning scarce or remarkable objects, often for their educational value and incidentally as a demonstration of the purchasing power or privilege of the individual collector. This emphasis on exclusivity and impact inspired the collecting of early, rare and unusual holograms.

For a larger group of consumers, the growing variety of holograms encouraged the acquisition of representative or exhaustive collections, in much the way that philatelists and numismatists collect stamps and coins. As discussed in Chapter 10, this activity accelerated as holograms became ubiquitous in advertising, popular art, packaging and commerce during the 1980s.

Jonathan Ross, for example, later a prominent collector, had been an early hologram entrepreneur in London (Figure 9.7). Influenced by the 1978 *Light Fantastic* show and new contacts, he set up a production company, SEE3 Holograms, and opened a commercial outlet, The Hologram Place. As Ross accumulated surplus stock and examples of the contemporary state of the art, his activities shifted towards exhibiting. Having had a sense from the beginning of being 'at the start of something', he began collecting in earnest and opened a permanent gallery of his pieces in 1998.[66]

For the affluent dilettante, rare holograms could reify the birth of a new idea or power. The first hologram marked a step in intellectual or technological history. A particular example could reveal scientific principles and capture audiences. But unlike cabinets of curiosities, holograms were seldom identifiably unique and rare. Most were created incrementally to explore or develop particular characteristics. As Emmett Leith ruefully recalled,

> People ask 'well, what was your first hologram, which is the first hologram?' And museums around the country and private collections and so on, people have enough of the 'very first hologram' around [like pieces of the real cross during the Middle Ages, when] you could find, in crypts and grottos, enough pieces of the real cross to start a lumber yard. And some of these holograms that people claim to be the original might be the thousandth.[67]

[65] Zellerbach, G. A., 'Selling holographic art: an analysis of past, present and future markets', *Proceedings of SPIE* 1600 (1991): 446–54.

[66] Ross, J. to SFJ, interview, 3 Apr 2003, London, SFJ collection.

[67] Leith, E. N. to SFJ, interview, 22 Jan 2003, Santa Clara, CA, SFJ collection.

Fig. 9.7 Jonathan Ross, Gallery 286, London, 1998 (photo by P. Boyd).

For collectors of holographic art, on the other hand, collections did not support a story of progressive revelation. Artistic concepts and expressions—the language of the imagery—valued a mature and stable statement, not iterative improvement. And, for the more democratic collector, the implicit theme was the illustration of variety and innovation. This was usually a straightforward story of growing technical diversity and cultural uptake. The tale of progress was wrapped up in it through the examples charting technological evolution and expanding markets. Such distinctive motivations and multiple voices inspired mixed story-telling via holograms, with collections showing off the first, the best, or the exemplary.[68]

This variety of motivations broadened the collecting of holograms to an extent unparalleled in, say, philately or sculpture. Collecting remained individualistic rather than a community activity having shared criteria or stable market values. Such collections were underlain by narratives of temporal and technical change, progress, nostalgia and exclusivity.

[68] Johnston, S. F., 'Attributing scientific and technical progress: the case of holography', *History and Technology* 21 (2005): 367–92.

10

Consuming Holograms

Burly security guards loomed over attendees at the International Conference on Optical Holography and its Applications. *In the corridors of what had once been the All-Union Lenin Museum, beautiful young women draped in 'holographic dresses'—glinting spectral folds of diffractive foil—attracted the delegates in ways that the hologram displays could not. In Ukraine, as in Asia, Europe and the Americas, commercial holography was now dominated by security applications. The muscle-power and glitz reflected the money and risks now at stake in the international hologram industry.*

On the walls of the convention centre, large reflection holograms celebrated the achievements of Ukrainian, Russian and Byelorussian holographers of the past decades. The subjects of their holograms and anecdotes were drawn largely from Soviet-era museums, lending the gravitas of heritage to what had once exemplified state-of-the-art technology. But most holograms during the noughties were being generated from digital files, not physical objects. Under the spotlights positioned over each booth, the attendees traded glimmering business cards, dazzling brochures and sheets of reflective foil stamped with intricate but low-cost holographic patterns. Their convoluted designs vied to show off technical skills for generating holograms that could protect, confuse or hide, conferring the mark of authenticity to banknotes, pharmaceuticals and luxury products.

Kiev, 2004

While audiences came and went, organizations were trying to turn holograms into a business. Exhibitions had revealed markets for new cottage industries: 'Lasers in Oxford Street – Holograms in the Home!'.[1]

Entrepreneurs and customers alike were captivated by holograms, but businesses struggled to develop new and competitive products and exploit that innovation effectively in the marketplace. Research companies lubricated by government contracts had been the first to investigate a commercial market for holograms from the

[1] The Hologram Place, 'Press release: Lasers in Oxford Street – Holograms in the Home!', London, 1978; G. Zellerbach collection. The firm operated a shop 'selling a range of commercial and limited edition holograms by British, American and European artists' alongside 'holographic jewellery which varies from kitsch to consummate craftsmanship', a service to advertisers 'to help design and create promotional holograms, for example, 3-D images on the Underground where you now see flat posters' and consulting to architects and interior designers 'in the realization of futuristic projects'.

late 1960s.[2] Their customers had been engineers like themselves, exploring potential applications.[3] Companies exploring holograms during the 1960s such as Conductron Corporation, IBM, Bell Telephone and CBS Laboratories had sought to stay at the front of the pack through basic research and by setting ambitious goals. They were confident that this would allow them to—as Dennis Gabor put it—*invent the future*.[4]

Dennis Gabor was, in fact, an interesting role model for entrepreneurs in holography. A Hungarian educated in Germany between the world wars, he had left with a few patents under his belt when Hitler came to power, and settled in England. Gabor proved to be a fertile inventor for his employer, British Thomson-Houston, working on everything from lamps to cinema technology to compact television tubes to nuclear fusion through his career. Physicist than electrical engineer, he was by all accounts a creative and deep thinker. And in his later years, after becoming a professor at Imperial College in London, he wrote about the role of technology in society—which is where that 'inventing the future' quotation came from.

But Gabor's commercial track record was poor. For someone so prolific in invention and so far-sighted about its social dimensions, he was an unsuccessful commercializer. His lamp designs of the 1930s could not compete in production cost with fluorescent tubes; his ideas for stereoscopic cinema of the 1940s worked only for small audiences; and his speech compression schemes of the 1950s were not pursued by their most likely sponsors, the military. In particular, Gabor failed to make any progress with what he called 'my favourite baby', holograms. He chided manufacturers of electron microscopes for their lack of interest.[5] By the late 1950s, even his closest collaborators were calling the invention a white elephant, and Gabor quietly downplayed the whole affair when he applied for a professorship.[6] And even after the explosion of interest during the 1960s, his exploratory corporate research as a consultant for CBS Lab failed to break into consumer markets.[7]

This pause in the expansion of holography was unexpected. It countered the pro-innovation bias of technologists and firms, which simplistically assessed new

[2] As noted by one analyst of the hologram industry, government-supported commercial activity—'institutional support, contract research, research grants, departmental purchases and commercialization assistance'—typical of late twentieth-century economies around the world, was a particular driver of hologram marketing (Kontnik, L. T., 'Governments underwrite holography industry', *Holography News*, 1 (4), 27 Dec 1987: 1).

[3] Lindgren, N., 'Search for a holography market', *Innovations* (1969): 16–27; Charnetski, C., 'The impact of holography on the consumer', presented at *Electronics and Aerospace Systems Conference Convention*, Washington DC, 1970.

[4] Gabor, D., *Inventing the Future* (London: Secker & Warburg, 1963). This phrase has also been an unofficial motto of the MIT Media Laboratory, which played an influential role in promoting holography from the 1980s; see Brand, S., *The Media Lab: Inventing the Future at MIT* (New York: Viking Penguin, 1987).

[5] Letter, D. Gabor to G. Rogers, 23 May 1958, Gabor papers, Science Museum MS 1014/11.

[6] Allibone, T. E., 'White and black elephants at Aldermaston', *Journal of Electronics and Control* 4 (1958): 179–92.

[7] Gabor envisaged holograms as 'an emotionally powerful method of teaching cultural history' (18 Jan 1972 letter Gabor to D. S. Strong, Imperial College Gabor MS/16), but generated no product ideas for CBS.

technology as universally valuable. A second wave of investment consequently sought new approaches. With a technology as unprecedented as holograms, there appeared to be unusual scope for defining new applications and new contexts of usage—ideas so original that they would help to define a 'social pull' to direct the 'technological push'.

There was limited success with corporate research. The most encouraging early applications were utilitarian spin-offs of militarily funded research on holographic optical elements (HOEs). These embody the functions of conventional optics—focusing, reflection and filtering—in a hologram. Military contractors Marconi and Pilkington, for example, used HOEs in designs for head-up displays (HUDs) in aircraft from the 1970s. In such viewing systems a hologram reflects the image of an instrument display to the eyes of the pilot, while allowing a clear view outside the cockpit. By combining several traditional optical functions in a single hologram, HUDs could be efficient and compact.[8] Similar advantages made holographic bar-code scanners popular in American supermarkets from the early1980s. Such scanners allowed a laser beam to be rapidly redirected in a complex scan pattern as it passed through a rotating hologram disk.[9] Yet neither of these profitable applications was much noticed by wider audiences: head-up displays, holographic or not, became a mundane element in the worlds of video-gaming from the 1990s (see Section 11.1), while the brief consumer interest surrounding supermarket scanners was the dancing red laser beam, rather than the holograms within.

During the following decade, a second wave of entrepreneurs explored industrial applications, sensed market trends, opened galleries and started cottage industries to produce holograms for popular audiences. The engineering newsletter *holosphere* [sic] (Figure 10.1), buoyed the nascent industries from 1972, until it was absorbed by the New York Museum of Holography in 1977 and broadened its remit to artists and entrepreneurs.[10]

During the 1980s, such efforts at commercialization enrolled a wider range of businesses and brought holograms to ever-larger audiences. From 1987, the glossy magazine *Holographics International*, run mainly by former physics students of Imperial College, London, provided a voice for the hologram industry, and Reconnaissance International

[8] See, for example, Harris, T. J., United States Office of Naval Research and International Business Machines Corporation, *Holographic Head-Up Display: Phase I* (Poughkeepsie, NY: International Business Machines Corporation, 1967); Boyle, D., 'The flying hologram-latest in head-up display systems', *Interavia (English Edition)* 36 (1981): 44–5.

[9] See, for example, Ikeda, H., S. Matsumoto and T. Inagaki, 'Hologram scanner for POS bar code symbol reader', *Fujitsu Scientific and Technical Journal* 15 (1979): 59–76; 'IBM supermarket scanner uses holographic optics', *Holosphere*, 9 (2), 1–2; Cato, R. T., 'Hand-held holographic scanner having highly visible locator beam', *IBM Technical Disclosure Bulletin* 27 (1984): 2021–2.

[10] See, for example, Harvey, T. W., 'Holotron reports results of holography market survey', *Holosphere*, 4 (4), April 1975: 1–3; 'Entrepreneurs urged to set up holographic centers', *Holosphere*, (6), 1976: 4, 8; 'Holotron licensing outlined by Cravens', *Holosphere*, 7 (9), Nov 1978: 3, 6; 'Shoppers get the picture: hologram marketers test the water', *Union*, 20 Jan 1977: 2.

holosphere, August 1975 7

holosphere, August 1976 7

Fig. 10.1 Engineers adapt to new markets (*Holosphere* 5 (8) (1975): 7 and *Holosphere* 7 (8) (1976): 7, courtesy of MIT Museum).

began a successful market analysis newsletter, *Holography News*.[11] And during the 1990s, *Holography Marketplace* appeared almost annually as a catalogue of techniques, hologram companies, products and services.[12] But for most participants, holography remained a frustrating and unprofitable business.

10.1 FITTING HOLOGRAMS TO CULTURE

> *I like abstract holograms. I've shown both kinds of works—abstract and literal—and the abstract works don't go over with the public. Mostly, I think it's because they're still at the stage where they want to see a hologram of a train.*[13]
>
> Gary Zellerbach, San Francisco, 1980

While collecting and exhibiting helped to preserve and valorize holograms, more regular applications remained elusive. Corporate explorations had not yielded a practical form of holographic television or cinema, and the rhetoric of 3-D media was beginning to fade as the pace of large-scale exhibitions declined. Despite the schools of holography, making holograms remained a specialist hobby. And the cottage industry that had been fostered by the institutional research, major exhibitions and hobbyist courses found too few customers. Holograms had spread from scientific to popular culture, but by the early 1980s, few holographers and entrepreneurs were able to eke out full-time livings by training enthusiasts or by selling holograms to consumers.[14]

The consumption of holograms confounded expectations. As a 1982 reviewer noted, 'Without a spotlight the photographic plate on which the image has been exposed will remain blank, but, this limitation apart, holograms are now perfectly serviceable household adornments.[15] Yet the impressive holograms that had generated throngs of exhibition visitors failed to sell in profitable numbers for display in the home. Practical limitations constrained home consumption and the holograms available seemed to generate an ephemeral visual appeal.

[11] See, for example, Lawrence, P. J., 'Retailing holography in the U.S.', *Holosphere*, 13 (3), Summer 1985: 8–9; Lancaster, I. M. and L. T. Kontnik, 'Market conditions for display holography in the USA and Europe', presented at *Practical Holography IV*, Los Angeles, CA, 1990; 'Into 1994: new developments in the marketplace', *Holography News*, 7 (10), Jul 1993: 88; Kontnik, L. T., 'Holography industry market survey and industry report: 1997', presented at *Sixth International Symposium on Display Holography*, Lake Forest, IL, 1997.

[12] The eighth and final edition appeared at the end of the century: Rhody, A. and F. Ross, *Holography Marketplace: the reference text and sourcebook for holography worldwide* (Berkeley, CA: Ross Books, 1999).

[13] Zellerbach, G. A. to J. Ross, interview, 1980, Los Angeles, CA, Ross collection.

[14] In California and London, two of the first commercial centres for display holography, participants surveyed in 1980 could identify no examples of full-time livelihoods (Zellerbach, G. A. to J. Ross, interview, 1980, Los Angeles, CA, Ross collection; Moore, L. to J. Ross, interview, 1980, Los Angeles, CA, Ross collection; McGrew, S. to J. Ross, interview, Dec 1980, Los Angeles, CA, Ross collection).

[15] McEwen, J., 'Art brought to light', *The Spectator*, 16 Apr 1982: 25.

Most purchasers discovered that holograms were frustratingly difficult to light well. The dramatic and flawless imagery of Leith–Upatnieks transmission holograms faded from popular memory because the cost and space needed for laser lighting had proved prohibitive. American and European legislation regarding laser safety had also shrouded cultural visibility through the 1970s; few laser-lit holograms escaped their labs for public display.[16] White-light holograms had their own lighting problems. Benton (rainbow transmission) holograms were similarly impracticable, because they, too, demanded careful lighting geometry and often had a limited viewing angle. And Denisyuk (reflection) holograms, despite painstaking attempts by English holographers to devise better processing formulas, tended to be dim even when lit by a compact light bulb at the correct angle.

One response was to devise systems incorporating the hologram, stand and light source (Figure 10.2); the idea evoked the schemes of the first stereoscopes over a century earlier. Another was to chemically bleach the hologram to reduce the absorption of light and produce a brighter image. Reflection holograms were best suited to this approach, but failed to become commercially popular. For consumers not accustomed to individually-lit artworks in the home, the need for a power source and all-too-obvious lamp looming from the frame was unappealing.[17] And bleached holograms, while brighter, were often veiled by scattered light. To make matters worse, even the carefully-made holographic artworks of the period could degrade unpredictably owing to the cocktail of experimental processing methods in vogue. Such practical problems had also hounded stereo views and early photography, which had proven less enduring than other forms of art and entertainment.

A much better visual alternative proved to be dichromated gelatine (DCG), a photosensitive medium that had been used for photographic and holographic purposes decades earlier and which generated a crystal-clear plate and bright reconstructed image (Figure 10.3).

But notoriously dependent on processing conditions and humidity, such 'dichromates' were ideally suited to artisanal cottage-industry techniques and could not readily be scaled up to mass production. Best suited to high-value art pieces, such holograms were typically recorded on glass plates and then chemically processed and encapsulated as a

[16] Consumer laser pointers were commercialized during the 1990s. Their growing availability, initially as an expensive lecture pointer but later as an optical toy for children, led to public concerns about eye safety during the early 2000s. On the other hand, the pointers finally made lasers cheap and familiar to the public. This had two consequences: a rapid decline in the mystique surrounding lasers—which had still been an important element in the attraction of the hologram shows of the 1980s—and an affordable way of making holograms at home, as discussed in Section 8.5.

[17] A later solution was Stephen Benton's 'edge-lit' hologram (c1990–4), which employed a prism to channel light into a glass-plate hologram at the optimal angle to reconstruct the three-dimensional image. The result was a framed hologram—either a rainbow type or holographic stereogram displaying a moving image—that incorporated the hidden light source. This was similar in format to the briefly popular 'digital frames' for displaying digital photos, although displaying a static three-dimensional image. See, for example, Benton, S. A., S. M. Birner and A. Shirakura, 'Edge-lit rainbow holograms', *Proceedings of SPIE* 1212 (1990): 149–57.

Fig. 10.2 Hologram with integrated stand, light source and power converter (4" × 5", silver halide on film; Third Dimension Ltd, 1988; M. Richardson collection).

glass-and-epoxy sandwich to prevent the ingress of humidity. Failure of the edge seal—a not-uncommon occurrence—would cause some or all of the holographic image to disappear. Through a variety of companies during the 1970s and eighties, entrepreneur Richard Rallison (1945–2010) drove commercial applications of dichromates. A more difficult problem was the subject matter of holograms. Few hologram makers of the period had access to a pulsed laser.[18] Nearly all holograms were of small inanimate objects, ranging

Fig. 10.3 Dichromate transmission hologram (Collector Richard Payne holding a diffraction grating composite hologram, *Mandala*, 4" × 5", created by Bill Reber in 1978; 2008 photo courtesy of J. Ross).

from the toy train in Leith's lab to the chess pieces typical of the late 1960s publicity, and the mundane knick-knacks of the first commercial holograms.

Cross (integral) holograms of the 1970s had addressed, and for a time solved, most of these problems. They were usually sold with a frame that supported the cylindrical film and an integral light source (an unfrosted light bulb). They could display a three-dimensional moving image of anything that could be filmed: outdoor scenes, computer-generated imagery and, most importantly, people. For museum and advertising use, they captured attention by their cutting-edge novelty and mystery. They were the most frequently reproduced examples of holograms through the 1970s, achieved wider publicity by being commissioned by Salvador Dali (*First Cylindric Chrono-Hologram. Portrait of Alice Cooper's Brain*, initially as a laser-lit prototype, 1973, and *Dali Painting Gala*, 1976), film makers (*Logan's Run,* Dir Michael Anderson, and *The Man Who Fell to Earth*, Dir Nicolas Roeg, both 1976) and point-of-sale advertisers (e.g. Sears & Roebuck).[19]

[18] Only a handful of pulsed lasers suitable for human subjects (i.e. having sufficient coherence for deep images) existed before 1980. Conductron's original pulsed ruby laser, developed by Larry Siebert in 1966, was inherited by McDonnell Douglas when the company was acquired, and then donated to the Smithsonian Institution and the Brookhaven National Laboratory in 1975 for a 'Center for Experimental Holography' that united art and science. Lloyd Cross developed a similar laser for KMS Industries in 1967, and Yuri Denisyuk and student Dmitri Staselko did the same in 1968. Other users of the period included groups headed by Ralph Wuerker at TRW (California), John Gates at the National Physical Laboratory (UK) and the All-Union Center for Photographic and Cinematographic Research (NIKFI) in Moscow, followed by Nick Phillips at Loughborough University and Hans Bjelkhagen at Holovision, Stockholm. See, for example, Bjelkhagen, H. I., 'Holographic portraits made by pulse lasers', *Leonardo* 25 (1992): 443–8.

[19] 'Shoppers get the picture: hologram marketers test the water', *Union*, 20 Jan 1977, 2.

Fig. 10.4 How to view a hologram, 1980s style (Jackson, P. *Through the Looking Glass: An exhibition of three-dimensional, floating images made with lasers* (Oregon Museum of Science and Industry, 1980), p. 6; illustration by Elizabeth Baeher; scan courtesy of G. Zellerbach).

Yet for home display, the cylindrical format demanded an unreasonable footprint (some 200 square inches), plus the much larger space to move around it (Figure 10.4). Mini- and half-cylinder versions were produced to reduce size and cost, but they also diminished the visual impact. A motorized rotating version could fit in the corner of a room, but the apparatus accommodated itself poorly to the average home. Motor noise, stray light and the need to position oneself to the image made the viewing experience reminiscent of slide shows of the previous generation. And as with the various varieties of flat holograms, they were prone to damage and deterioration after years of storage. While many thousands were sold, few appear to have survived for eBay auctions four decades later.

While these creative and entrepreneurial threads intermingled, another application was explored: holographic portraits (Figures 10.5 and 10.6). The pulsed ruby lasers developed during the 1960s were gradually joined by newer, commercially-available systems. By 1980, though, entrepreneurs saw pulsed lasers not as hologram hobbyists' flashbulbs, but rather as a tool for professional portraits. Peter Nicholson, an early associate of Cross, had used the old Conductron laser during the late seventies to create such portraits; Nick Phillips did the same for the 1978 *Light Fantastic* exhibition; artist Margaret Benyon had created pulsed hologram artworks from 1981 with physicist John Webster, and from the mid-1980s the students of the Holography Unit of the Royal College of Art in London explored the medium further.[20] By 1980, pulsed portraits were being offered commercially for a few thousand dollars, with copies at $700.[21]

[20] Benyon, M., 'Pulsed holographic art practice', presented at *Practical Holography*, Los Angeles, CA, 1986.
[21] Lasergruppen Holovision AB (Stockholm, Sweden) price list, c1980; G. Zellerbach collection.

Fig. 10.5 Pulsed laser portrait holograms ('Boy George' by Edwina Orr and David Traynor, 1985; 'Lucy in a Tin Hat' by Patrick Boyd, 1989; 'Herbie Hancock' by Anna-Maria Christakis, Musée de l'Holographie, Paris 1986. Pulsed laser reflection holograms, silver halide on glass, each approx. 43 × 32 cm². Photo from 2009 Buckinghamshire County Museum exhibition, Aylesbury UK, courtesy of J. Ross).

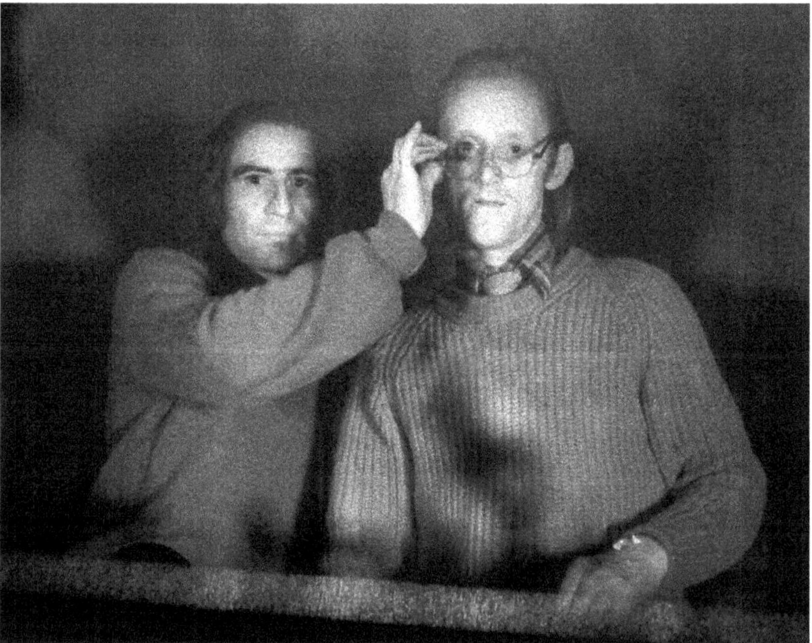

Fig. 10.6 Holographic 'selfies' (Anton Furst and Nick Phillips, Holoco Ltd, as creators and subjects 1977. Pulsed laser transmission hologram, silver halide on glass; photo courtesy of T. Dalenius, 2013).

Through the 1980s and 1990s, in particular, these activities sampled distinct cultural strands. Holographic artists expanded their repertoire from still-lifes, abstracts and mixed media to include the human figure. Holographers such as Anna Maria Nicolson and Harriet Casdin-Silver were commissioned to do ethnographic series that captured the diversity of populations, but also the famous or renowned. Art school holographers represented their times, recording their peers to capture images typical of the punk and New Modern periods. Portraitists sought to capture the great and the good in poses reminiscent of early twentieth-century formality. Ex-President Ronald Reagan (1991) and Queen Elizabeth II (2003) sat for holographic portraits and—at the other end of the cultural spectrum—*Playboy* models (2001) were the subjects of a series of commercial pieces.[22]

Pulsed holograms also converged with popular culture in another sense: they occasionally mirrored hobby photography by capturing either candid shots or self-portraits (Figure 10.6). Such examples challenged popular conventions surrounding the medium. Prior holograms, at least those seen publicly, had been carefully posed and purposeful; these examples revealed qualities of playfulness and casual spontaneity.

On a deeper level, these unscripted products integrated popular notions of holograms. Holographic portraits, like other holograms, recorded more than their creators usually recognized. Stray laser light bouncing off equipment, walls, and bystanders could reveal the wider environment (Figure 10.7). Holograms embodied the hidden and shadowy, as well as the brightly lit. As they awed their viewers, they also encouraged exploration of their depths. Rediscovering such captured scenes evoked wonder in ways that formal portraits could not.

Yet portraits, too, failed to attract many consumers. Viewers of the first exhibited pulsed-laser portraits were taken aback by their frozen realism. The single-colour light produced a monochromatic image that resembled the unnatural tonality of early orthochromatic black-and-white emulsions. Artists who adopted pulsed lasers employed them most successfully for figure studies rather than facial depictions.[23]

Affluent patrons for personal portraits were equally unimpressed by the impossibility of retouching or editing their captured likenesses. As each laser pulse took place in a

[22] Bjelkhagen, H. I., R. E. Deem, J. Landry, M. E. Marhic and F. D. Unterseher, 'Holographic portrait of Ronald Reagan', presented at *International Symposium on Display Holography*, Lake Forest, IL, 1991; 'Holographic playmates', *Holography News*, 15 (2), Mar 2001: 4. The Reagan and playmate portraits were made with a pulsed laser; the Queen Elizabeth portrait, recorded by Rob Munday, Chris Levine and Jeffrey Robb, was a rainbow transmission holographic stereogram created from 205 individual digital photographs. Celebrity holograms included a series sponsored by the New York School of Holography, *New Yorkers Who Make a Difference*, in which Ana Maria Nicholson recorded celebrities such as Gloria Steinem, David Byrne, Arthur Schlesinger, George Plimpton and Walter Cronkite; Richmond Holographic Studio captured Boy George; Martin Richardson recorded David Bowie, Martin Scorsese, Peter Blake and others; Jason Sapan's integral holograms included portraits of Edward Heath, Billy Idol, Isaac Asimov and Andy Warhol.

[23] Only a handful of pulsed laser portrait studios appeared, notably Conductron (Michigan, late 1960s), McDonnell Douglas (Missouri, early 1970s), the Museum of Holography (New York, 1980s), Musée de l'Holographie (Paris, 1980s), Laser Reflections (California, 1990s), Spatial Imaging (England, 1990s), Holicon (Illinois, 1990s) and VVC (Moscow, 1990s).

Fig. 10.7 (a) Martin Scorsese holographic portrait and (b) dimly-lit witness Graham Tunnedine on the side-lines, incidentally recorded by the stray light of the laser pulse (Richardson, M., portrait of Martin Scorsese, pulsed laser transmission hologram, silver halide on glass, 1998; photos by S. Johnston, 2013. See also Richardson, M., 'Hidden images of holography: wavefront reconstruction of abnormalities within pulsed holographic recording', *Proceedings of SPIE* 9006, *Practical Holography XXVIII: Materials and Applications* (2014)).

darkened room, the subject's irises were likely to be large, the expression fixed or unprepared. The first lasers employed for pulsed-portraits also recorded skin imperfections and pallid complexion, a result of the transparency of the skin to the deep red light of the ruby laser.[24]

Holographers and entrepreneurs began to more actively try to identify potential consumers, and to adapt holograms to their desires. In its way, this was belated recognition of a surprising failure. Holograms, so ardently promoted from the sixties as an expression of the future, had not yet created it. As noted by Kenneth Haines, an energetic and innovative contributor to the field for over forty years, 'We were trying to exploit holography—[but we] didn't know what the hell to do with the thing'.[25]

10.2 CANNING THE UNCANNY

The link between modernity and visual surprise had been noted by Walter Benjamin in 1939 in relation to cinema. The evocation of the sublime—combining both excitement and a degree of discomfort or fear—was noted by contemporary viewers, too. When Maxim Gorky (1868–1936) described a screening of the Lumière Brothers' *The Arrival of a Train* in 1896, he contrasted the looming physical threat with its shadowy unreality. This sense of the uncanny—the melding of the familiar with the unsettling—was, for Gorky, the primary emotion evoked by early cinema. As film theorist Tom Gunning put it, 'it is the incredible nature of the illusion itself that renders the viewer speechless'. Gunning has argued convincingly that a basic aesthetic of early cinema was this series of visual shocks, describing it as 'the climax of a period of intense development in visual entertainments'.[26]

The same was true for audiences encountering holograms. They were not always naïve observers; just as often, spectators were intrigued by the incongruity of intense reality with immateriality. Producers discovered that the appeal of holograms lay, initially at least, in their ability to generate uncanny experiences for critical audiences.[27]

But the history of commercial holography shows a steady movement away from such disturbing experiences towards gentler surprise, novelty and, eventually, utility. In the quarter-century between 1970 and 1995, the hologram was tamed. For one exhibition, organizer Jonathan Ross sought to play down the element of spectacle: rather than 'employing holography in a gimmicky way . . . these are not holograms that will make

[24] Nicholson, A. M. to SFJ, interview, 21 Jan 2003, Santa Clara, CA, SFJ collection; Munday, R. to SFJ, interview, 31 Mar 2004, Richmond, UK, SFJ collection.

[25] Haines, K. to SFJ, interview, 21 Jan 2003, Santa Clara, CA, SFJ collection.

[26] Gunning, T., 'An aesthetic of astonishment: the (in)credulous spectator', *Art and Text* 34 (1989): 31–45, quotation p. 35; Gunning, T., 'The cinema of attraction: early film, its spectator, and the Avant-Garde', *Wide Angle* 8 (1986): 63–70.

[27] See Royle, N., *The Uncanny* (Manchester: Manchester University Press, 2003).

you gasp at their three dimensionality or wonder at their verisimilitude (although perhaps you may marvel at their beauty)'.[28]

As entrepreneurs founded cottage industries to manufacture holograms and opened sales outlets, they discovered that uncanniness was short-lived. Fine-art holograms could intrigue and bemuse audiences in gallery settings, but few were purchased for home display. Consumer holograms—seldom larger than four by five inches—scaled-down the experience accordingly.

Gary Zellerbach (b. 1952), founder of the Holos Gallery in the Haight-Ashbury district of San Francisco, one of the first hologram galleries and distribution businesses, documented this changing aesthetic. From the early 1980s, commercial holograms were available from a growing collection of hologram manufacturers in the USA and Western Europe. Stock imagery on glass plates or Multiplex cylinders could be had wholesale for between a few hundred and a few thousand dollars, but even without the two-fold markup for retail prices found few buyers.[29] Zellerbach moved progressively from limited edition wall hangings towards cheaper mass-market holographic curios. Accounting for about 1% of the gallery's income in 1980, art sales were replaced almost entirely by kitsch ornaments a decade later, with dichromate pendants and small rainbow holograms on film selling for around $10 or less.[30]

Hologram retailers struggled to build a consumer market by broadening their product ranges, and by allying themselves with science museums and technology centres to capture national, rather than local, audiences.[31] Such affiliations constrained the subject content further. Established suppliers like Edmund Scientific, targeting technology enthusiasts and schools, tended to offer holograms that illustrated either striking visual effects or educational value.

Holographic novelties proliferated during the 1980s, when visual surprise could attract waves of new audiences. The 'Holodisk', for example, was a rotating animated hologram which could display a titillating strip-tease or comic book characters, for adult and youthful audiences, respectively.[32] For other sellers, fashion items such as holographic medallions and earrings extended the market possibilities, but also sacrificed the impact of the imagery: the effectiveness of the jewellery depended on how it was lit.[33]

[28] Ross, J., 'Press release: Jonathan Ross presents LANDSCAPES & METAMORPHOSES: an exhibition of work by Jeffrey Robb,' London, 1993; G. Zellerbach collection.

[29] Invoices and wholesale price lists, Advanced Dimensional Displays (Van Nuys CA, Feb 1988); Advanced Holographics Ltd (London, 22 Oct 1987); AP Holographie (Paris, Jan 1985); Dichromate Inc (1980); Holex Corp (Philadelphia, Summer 1980); Holos Gallery (San Francisco, Jul 1990); Lasergruppen Holovision AB (Stockholm, 1980); Multiplex Company (San Francisco, 1978); New York Holographic Laboratories (1980); Richmond Holographic Studios (London, c1985); Sapan Engineering Co (New York, 1978); Third Dimension Ltd (London, 1988); White Light Works (Woodland Hills CA, c1980); all from the G. Zellerbach collection.

[30] Zellerbach, G. A. to S. Johnston, telephone interview, 12 Jun 2014, SFJ collection.

[31] 'It's Spring, and holo galleries are popping up from coast to coast', *Holosphere*, 8 (3), Mar 1979: 1.

[32] 'Lovely Rita', reflection H1/H2 hologram, silver halide on film, designed and manufactured by C. Outwater and C. Newswanger of Advanced Dimensional Displays (ADD), 1987. The original products sold for about $20. Similar holograms depicting comic book characters were produced as late as 1995.

[33] 'Holograms as giftware: the commercialization of artistic holography', *Holography Marketplace* 8 (1999): 37–43. The editor of *Holographics International* later complained that holographic jewellery had become 'twee and uninteresting', was poorly made and devoid of style and cachet (Bains, S., 'Editorial', *Holographics International* 1 (1989): 4).

Fig. 10.8 Arcade game employing a Multiplex hologram (Midway Manufacturing Co., 1976; S. Johnston collection).

Manufacturers explored the market for children's games that used holograms. The earliest were arcade games in which a rotating cylindrical hologram made by the Multiplex Company provided animation for fighting games that included Midway's *Top Gun* (Figure 10.8) and Kasco's *Gun Smoke, Samurai* and *Bank Robber* arcade games (1975–77).[34]

[34] Few 'hologram' games incorporated actual holograms. Examples include *Time Traveller* (1990), patented by Dentsu Inc., Japan and developed by the company With Design In Mind, and Sega's *Holosseum* (1992), another fighting game. *Time Traveller* and *Holosseum* used a mirror to project images from a videodisk television display to viewers' eyes.

Atari, the electronic game manufacturer, trialled a hand-held electronic toy in 1981 (Figure 10.9). Cosmos, developed by Steve McGrew and Ken Haines, incorporated interchangeable holograms as backgrounds. This 'whole new dimension to electronic computer games' nevertheless left trade show audiences largely unimpressed, and was never marketed.[35] Atari had been developing a second-generation handheld version dubbed Spector, but this, too, failed to reach market. Such games straddled an awkward divide between early videogames, on the one hand, and optical novelties, on the other. Roger

holosphere

Atari Unveils Cosmos, A Holographic Game System

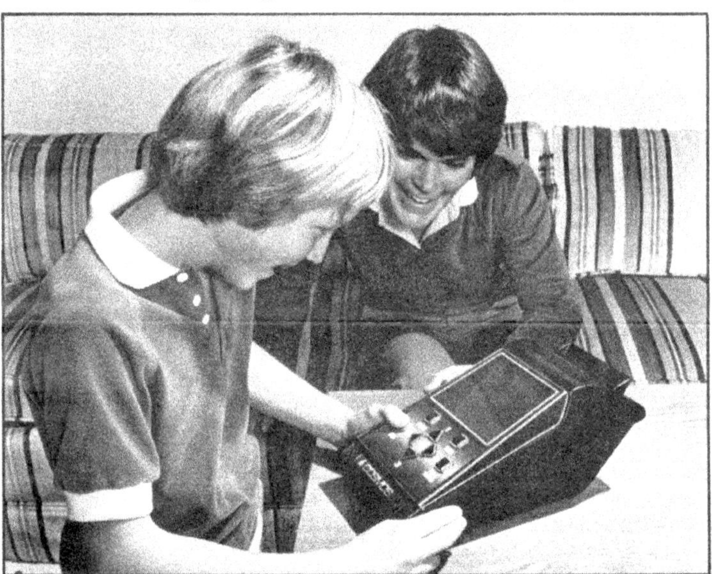

"Cosmos – The Third Dimension – is a revolutionary new game system that combines breathtaking 3 Dimensional Holoptic images, sophisticated LED game play and the most innovative sound effects available in an electronic game" (photo and caption courtesy Atari, Inc.).

Fig. 10.9 Cosmos game system, 1981 ('Atari Cosmos game system' *Holosphere*, 10 (3), Mar 1981: 1).

[35] 'Atari Cosmos game system', *Holosphere* 10 (1981): 1; McGrew, S., 'Mass produced holograms for the entertainment industry', *Proceedings of SPIE* 391 (1983): 19–20.

Hector, one of Atari's designers, called Cosmos 'a cutting-edge exotic electronic toy, not a low-end video game console'.[36]

Such consumer products exploited hologram technology to mine new markets. The holograms of the late 1970s and early 1980s had limited manufacturing options. Cottage-industry holograms had begun as handmade pieces recorded on glass plates or photographic film, but a handful of firms developed manufacturing processes to mass-produce consumer holograms. Stories about hologram businesses could entertain audiences as much as the holograms themselves could: they revived the story of inevitable progression of technology spiced with intriguing investment opportunities and consumer entertainment to come.[37] In Britain, Applied Holographics touted an automated 'holocopier' to record holograms on conventional silver halide emulsions using a pulsed laser.[38] Supported by investment from merchant bankers and government sponsorship, the company collaborated during the late 1980s with photographic firm Ilford Ltd to manufacture holograms for the toy industry. As adhesive reflectors on *Dungeons and Dragons* games, Tonka action figures and cereal prizes, such products represented the most inexpensive and widely distributed holograms of the period based on 1960s technologies.[39]

Yet another medium that had its own distinct pattern of distribution and usage was photopolymer. Photopolymers used in holography are transparent long-chain molecular materials that cross link when exposed to short-wavelength light. Unlike DCG, which had been nurtured by entrepreneurs and cottage industry after corporate interest waned, photopolymers began as, and remained, a product of large companies. From the late 1980s, the Polaroid Corporation marketed a grainless photopolymer recording material requiring a carefully controlled wet process. Du Pont Inc developed photopolymer materials in its Delaware laboratories in 1975, and eventually introduced commercial photopolymer materials developed by ultraviolet exposure in 1989. Early photopolymer holograms were demonstration pieces distributed at conferences or for use on catalogue covers as sales promotions, although large production runs for corporate customers such as Disney and DC Comics began to appear through the 1990s.[40]

Such explorations sought inexpensive processes that could record the fine fringe pattern of the hologram onto either a material (via variations in transparency, creating an 'amplitude hologram', or in density, creating a 'phase hologram') or onto a surface (creating a 'surface relief hologram'). Among the more imaginative notions was holographic candy and chocolate: transparent suckers and reflective chocolate could be moulded or imprinted to carry holographic images. Dimensional Foods, established by MIT alumnus Eric Begleiter,

[36] Hector, R., 'Interview with Stilphen', http://www.2600connection.com/interviews/roger_hector/interview_roger_hector.html, accessed 15 Aug 2014.

[37] Kay, W., 'Why laser men are beaming', *Sunday Times*, 17 Apr 1983, 54; Nuttall, G., 'Made in Britain: the new millionaires who beat the slump', *Sunday Times Magazine*, 17 Apr 1983, 31–6; photo p. 31 by Iain McKell.

[38] Silver halide is the chemistry of conventional photographic emulsions, and the most popular process since its introduction in the 1840s.

[39] Pepper, A., 'Expansion of holographic film products', *Holosphere*, 15 (1), Spring 1987: 12–3.

[40] The covers of successive *Holography Marketplace* catalogue incorporated several impressive examples, one in full colour.

found such edible holograms difficult to market. Focus groups revealed that consumers were often concerned that the bright diffractive colours of the candies were due to artificial dyes, which had become the focus of food safety campaigners from the 1980s. Others worried that the sparkling colours indicated unhealthily high sugar content. Begleiter's attempts to win over consumers with explanations of 'tiny prisms' were unsuccessful.[41]

But in other guises, such embossed holograms transformed business profitability and, with it, cultural engagement by mass audiences. Embossed holograms had been invented in the late 1960s, were developed for consumer applications during the 1970s, and flowered a decade later.

Originated by Robert Bartolini and colleagues at RCA Corporation, the concept of embossing was the basis of a videoplayer (SelectaVision) that would use a laser to read holograms recorded on a plastic tape. The 'holotapes' themselves would be manufactured literally by stamping the holographic patterns into soft transparent vinyl. This low-cost replicating process was not brought to market, however. The proof of concept models, planned for sale in late 1972, were set aside in favour of magnetic recording, which would allow consumers not only to play back, but to make their own video recordings.[42]

An independent inventor, Mike Foster, developed a novel method of transferring holographic information from conventional photographically-recorded holograms to nickel embossing shims, which could then be used to stamp holograms onto plastic. As early as 1974 the potential of the process was being publicized for holographic gift-wrapping, clothing, and more:

> There is no doubt that it is going to have an early impact on the record industry. Holograms can be embossed directly onto records by using many of the same techniques already used for pressing the sound grooves into the vinyl. Grand Funk, a rock band, will soon put out a long-playing record made out of defractory [sic] vinyl, simply as something nice to look at. In the center of the record, a giant finger will be pointing up and out in 3-D . . .
>
> No one could accuse Mattel Toys of creating products having much to do with the jet set, yet the company has already leased an option on Mike Foster's embossing method and is planning to market a new line of holographic toys in the near future. Included will be a small, holographic movie-viewing device. Mattel also plans to manufacture games building on the principle that two people looking at the same hologram from different viewpoints see different things. The firm also sees a big market for holographic playing cards featuring everything from baseball players to antique cars.[43]

[41] Begleiter, E., 'Edible holography: the application of holographic techniques to food processing', presented at *Practical Holography V*, San Jose, CA, 1991; Begleiter, E. to SFJ, interview, 10 Jul 2003, Cambridge, MA, SFJ collection. The idea was revived by Morphotonix, Switzerland, in 2014.

[42] Bartolini, R., W. Hannan, D. Karlsons and M. Luric, 'Embossed hologram motion pictures for television playback', *Applied Optics* 9 (1970): 2283–90; Bartolini, R. A., J. Bordogna and D. Karlsons, 'Recording considerations for RCA Holotape', *RCA Review* 33 (1972): 170–205; Bartolini, R. A., 'Holographic motion pictures for TV playback', presented at *1973 IEEE/OSA Conference on Laser Engineering and Applications Digest of Technical Papers*, New York, 1973; Groskinski, H., 'Cassette TV: the good revolution', *Life*, 69 (16), 16 Oct 1970: 46–53.

[43] Rolfe, L. and N. Lennon, 'Holography – new light on a new dimension', *Delta Airlines Sky*, Dec 1974: 23–29.

Foster founded Spectratek in 1980 to produce holographic foils for packaging, and produced holographically-stamped vinyl recordings for Styx and Split Enz releases the same year.[44]

A second seminal innovator was Steve McGrew (b. 1945), who envisaged embossed holograms as inexpensive graphic arts medium. In 1977 he formulated replication processes based on ultraviolet and thermal curing of polymers and, along with them, applications of how they could be used. He imagined holograms cheap enough to be used anywhere: car windows, greeting cards and book jackets, if treated with a reflecting layer.[45] More influentially, McGrew conceived such reflective foils for a particular kind of hologram: relatively flat image-plane rainbow holograms. His so-called 2D–3D embossed holograms were synthesized from flat artwork—making them appealing to commercial graphic artists—and reconstructed an image in two planes, one a few millimetres behind the surface of the hologram and another at the surface itself. The advantage of this depth restriction was greater ease of lighting and viewing the bright images. His crude but colourful *Holographic Sunrise*, a six-inch square embossed souvenir image, sold quickly to consumers and encouraged McGrew to expand production and marketing. Supported by a rigid cardboard or plastic backing, the incomparably bright holograms proved inexpensive and appealing to customers. The shallow imagery reached wide audiences in 1982 through promotional stickers used in a marketing campaign for the movie *ET*. Unusually in the history of optical novelties, embossed holograms reached children's markets at the same time as those for adults.

A third key innovator was Kenneth Haines (b. 1938), one of the first generation of engineers at Willow Run Labs developing holography and a peripatetic contributor to the technology in a variety of firms. During the late 1970s at Eidetic Images Inc, Haines developed an improved version of Lloyd Cross's cylindrical multiplex hologram and, seeking an inexpensive method of reproducing them, he developed a pressure-moulding process. Haines was a key contributor during the 1980s, responsible for embossed hologram covers on *National Geographic* magazine and numerous series of sports cards.[46]

Embossing enrolled the techniques, companies and customers of the traditional printing industries, adding a network of established businesses to support holographers.[47] Via the applications pursued by Foster, McGrew and Haines, this new breed of hologram created strong impact.

[44] Stamping machines to manufacture embossed holograms were, in fact, similar to those used to manufacture compact disks, which had first been marketed by Philips in 1981.

[45] McGrew, S. to J. Ross, interview, Dec 1980, Los Angeles, Ross collection.

[46] Covers included an embossed hologram of an American eagle model (Mar 1984), a primitive human skull (Nov 1985) and an exploding crystal globe (Dec 1988). Haines, K., 'Development of embossed holograms', *Proceedings of SPIE* 2652 (1996): 45–52.

[47] 'Packaging and design sells: Holography News interviews Barc Thompson', *Holography News*, 7 (1), 1993: 3–4; 'Holographic Image: a Virginia-based company is keeping busy with the growing popularity of holographic materials because it supplies the embossed materials as well as the application equipment', *Paper, Film and Foil Converter* 68 (1994): 90; Clark, A., 'Making an impact: holographic applications for print, and looks at developments in foiling and embossing', *The British Printer* 109 (1996): 34.

Fig. 10.10 Periodical with reflective embossed holograms (*Holography Marketplace*, 4th ed. (1993); S. Johnston photo). Compare with Figure 3.2.

Embossed reflective holograms in magazines proved to be a short-lived fashion in publishing (Figure 10.10), however. During the mid-1980s to mid-1990s, magazine covers and advertisements flirted with the aluminized rainbow images. Among the first holographic covers was a 1983 issue of the British magazine *Amateur Photographer*.[48] Graphic designers experimented with embossed elements ranging from small insets to complement conventional printed images, to entire holographic front and back covers formed from an embossed sheet.[49] The commercial experiments led to a boom in other types of 3-D cover, notably anaglyph photographs (which required coloured glasses) and lenticular images (described in Appendix).

The decline of magazine holograms was at least partly due to the difficulty of viewing them. Unless the light source was small and positioned at the design angle, the holographic image would be blurred and dim. The flexible pages of the publication distorted the reflective hologram and changed its angle with respect to the light source, giving a blotchy reconstruction. Viewers found themselves tilting and rotating their magazines to find the unique position that yielded a sharp, colourful image. Invoking irritation more often than fascination, the medium was an unpromising advertising medium. As Emmett Leith, one of the key inventors of holography, observed,

[48] *Amateur Photographer*, 167 (26), 25 Jun. 1983: cover. Other widely seen covers included *Sports Illustrated*, with a holographic portrait of Michael Jordan (Dec 1991) and the British magazine *New Scientist* (Feb 1988).

[49] The December 1988 *tour de force*, in which the front cover, spine and back cover (an advertisement for McDonald's hamburgers) were a single embossed hologram sheet, proved to be a project fiasco costing *National Geographic* over 2 million dollars (Haines, K. to SFJ, interview, 21 Jan 2003, Santa Clara, CA, SFJ collection).

> Along the way, as one by one the needed elements materialized, holographers always said that now only one more element was needed for the success of display holography, and when it came, there was always just one more element needed. Have we now reached the end of the long sequence of requirements? Not quite yet. To be really successful, holography must be much more forgiving of the lighting and viewing conditions.[50]

In other cultural contexts, these limitations were not so noticeable. Children's collector cards, sticker-books and plastic toys of the period were dominated by the bright and colourful images.[51] Beginning with Zebra Books in 1985, which added distinctive 2D–3D holographic logos on the covers of its romance novels, other publishers adopted stamped holographic cover images during the late 1980s. Greetings cards—one of the first applications conceived for embossed holograms—also appeared commercially at this time.[52] As suggested by their descending position towards the lower end of the publishing market, holograms in the graphic arts were recognized as a short-lived fad. The technique and its market applications both created shallow images.

As a wider range of conventional printers were tempted to adopt simpler holographic imagery, abstractly-patterned cereal boxes, toothpastes and product labels attracted consumers' attention by their random sparkling in well-lit supermarket aisles.[53] SmithKline Beecham's Aquafresh Whitening toothpaste, launched in 1994, pioneered the market. And, as the product packaging market became mature and saturated, holographically-patterned gift-wrapping foils added new excitement to presents from the late 1990s.[54]

Holographic postage stamps—bearing either a reflective diffractive foil or a transparent diffractive foil over conventional printing—were introduced by Austria in 1988, and by 2015 some eighty other countries had followed with commemorative and souvenir stamps. As with other embossed holograms, holographic stamps had a restricted period of popularity, peaking during celebrations of the millennium.[55] From 1999 coins, too, were imprinted with holographic patterns, principally by the Royal Canadian Mint. Like most commemorative holographic stamps, proof gold and silver coins bearing holograms were intended for collectors, seldom circulated and hence rarely seen.

More recently, holographic flakes have been popular in nail glitter, cosmetics, fishing lures and clothing. In each market, three-dimensionality and then imagery were successively compromised and abandoned, leaving only movement and colour behind. The result, to paraphrase Emmett Leith's description of how his first holograms focused images, was 'kind of like a grin without a [Cheshire] cat'.[56]

[50] Leith, E. N., 'Reflections on display holography', *L.A.S.E.R. News*, 10 (4), 1990: 1.

[51] Sadowski, P., 'Holographic child's play', *Holosphere*, 15 (1), Spring 1987: 17.

[52] American Bank Note (ABN) purchased licenses from Holosonics in 1980 specifically for the greeting cards market.

[53] Ralston-Purina introduced Ghostbusters cereal boxes with a series of three embossed holograms in 1986, reportedly tripling sales (Rhody, A. and F. Ross (eds.), *Holography Marketplace 7th Edition* (Berkeley, CA: Ross Books, 1998), p. 40).

[54] Editor's Choice, 'Holography ideas', *Potentials in Marketing*, Sep 1991: 14.

[55] Bjelkhagen, H. I., *Holography and Philately: Postage Stamps with Holograms* (Dartford, UK: XLibrisUK, 2014).

[56] Leith, E. to S. F. Johnston, interview, 22 Jan 2003, Santa Clara, CA, SFJ collection.

These decorative graphics arts defined an era. Embossed holograms altered contemporary culture, giving the 1980s a visual expression as iconic as the finned automobiles of the 1950s and skyscrapers of the 1920s. While embossed holograms contributed to the zeitgeist of the late twentieth century, their appeal was broadly recognized by their producers as ephemeral. Trading novelty for aesthetics, they entered popular culture as commercial graphics and a children's fad, ending up as subliminal visual prompts in daily life. As Walter Benjamin might have observed, as holograms became ubiquitous and viewable under different lighting, their aura was lost. The special environments and practices surrounding the act of viewing was replaced by mass consumption.[57] Summarized by entrepreneur and collector Jonathan Ross, 'they were kitsch and crap and "shiny shit" in common terminology, but it seemed a way to make money out of it'.[58]

10.3 SYMBOLS OF SECURITY

Mass production of holograms through embossing processes multiplied their numbers from tens of thousands in the mid-1970s to tens of millions fifteen years later. But the 'killer application' for embossed holograms was anti-counterfeiting.

Holograms had first been conceived as encoding and decoding devices by Emmett Leith and Juris Upatnieks in 1963.[59] Acting as an optical diffuser that could jumble and subsequently recover an image, it was a concept familiar to their generation of schoolchildren as cereal-box secret decoder rings.[60]

Money transactions proved a better application. A general purpose 'charge card' had been introduced by American Express in 1950, but similar ideas were adopted in other countries only from the late 1960s. Verifying the authenticity of credit cards was not a new idea. As with paper money, firms had a long-standing interest in detecting counterfeits. In early credit cards, the stamped plastic number and an appended signature sample were novel enough to make counterfeiting difficult. From 1971, some credit cards incorporated magnetic stripes to add an extra dimension of complexity to discourage fraudulent manufacture, but these had the drawback of demanding more expensive

[57] Benjamin, W. and A. Underwood (transl.), *The Work of Art in the Age of Mechanical Reproduction* (New York: Penguin, 2008).

[58] Ross, J. to SFJ, interview, 3 Apr 2003, London, SFJ collection. The label was widely used by commercial holographers, e.g. Kontnik, L. T. to SFJ, interview, 20 Nov 2003, Vancouver, SFJ collection and Munday, R. to SFJ, interview, 31 Mar 2004, Richmond, UK, SFJ collection.

[59] Leith, E. N., 'Proposal for Applications of the Wavefront Reconstruction Technique', Grant proposal ORA-63-1255-PB1, Institute of Science and Technology, University of Michigan to US Army Research Office, Durham, NC, 28 May 1963.

[60] Such rings had been a feature of children's radio shows, and sponsor promotions, from the 1930s, beginning with *Little Orphan Annie* and the Ovaltine Company. Many were alphabetic substitution ciphers, but some included transparent templates or colour filters to mask or reveal printed messages on product packaging.

magnetic readers at the point of sale. As early as 1970, holograms were being considered for assuring genuine transactions.[61]

But embossed holograms were useful as security devices in a more obvious way. Their complicated optical behaviours—reflecting spectral colours and producing multiple images or animated motion as viewing angle changed—made them much more difficult to copy than conventional printed labels. The soft aluminized plastic was also easily distorted or obliterated, making mechanical tampering obvious. Such holograms would be viewable by both the merchant and the customer, and the complexity of the image would require no further equipment or cost to either.[62]

The director of American Bank Note (ABN), Ed Weitzen, conceived the application and his company acquired a license from Holosonics, then the principal patent holder on holograms, in 1980. ABN exploited the concept through a series of acquisitions—first acquiring Ken Haines' company, Eidetic Images, which developed the application, and then manufacturing them through another acquisition, the embossing company Old Dominion Foils. The combination of expertise spawned a new company, American Bank Note Holographics (ABNH). Weitzen proposed credit card holograms to MasterCard; by the end of 1982, ABNH was trialling the process, and Mastercard began distributing the new cards to its customers the following year. In 1985, the ABNH technology was introduced in Visa credit cards, too. While small and scarcely noticed, embossed holograms became almost universal on credit cards within a decade. These two credit card companies alone had over 400 million cards in circulation by the turn of the twenty-first century.

The second wave of holographic security products were employed for other monetary documents—debit cards, cheque guarantee cards, travellers' cheques, bonds, and high-denomination banknotes—and became ubiquitous during the 1990s owing to the marketing of a handful of large security producers such as De La Rue Holographics in Basingstoke, UK.[63] Governments, too, profited: holographic tax stamps ('banderols') were adopted as official seals on alcohol and cigarette packages from the mid-1990s. These were soon followed by a third wave: identity documents without direct monetary value, such as passports, visas, drivers' licenses and employee cards. And the fourth wave was the holographic marking of high-value commodities: highly-taxed spirits and cigarettes, expensive pharmaceuticals and designer brands. Glaxo was the first pharmaceutical company to protect its popular Zantac drug with a holographic authenticity label in 1988. A conservative estimate of the global population of such holograms suggests that several billion had been produced by 2000.

[61] Christy, E. H. and K. K. Sutherlin, 'Validating credit cards using holography', *IEEE International Convention Digest* 4 (1970): 338–9; Schuenzel, E. C. and R. L. Moore, 'Credit card system', *IBM Technical Disclosure Bulletin*, 13 (1970): 176–7.

[62] Subsequent security schemes, such as chip-and-pin (personal identity number) systems, rely on electronic communications between the merchant and credit card company to authenticate transactions.

[63] See, for example, Bloom Murray, T., *The Brotherhood of Money: the Secret World of Bank Note Printers* (Port Clinton, OH: BNR Press, 1983).

Economic judgements were clear-cut: where art holography had failed, holographic authentication was a booming success.[64]

The new high-volume, high-income industry (at least for the handful of security firms dominating the market) evolved rapidly. Commercial competition became intense, with counterfeiting operations paralleling them; the sophistication of security holograms rose accordingly.

The first high-profile case was the 1991 counterfeiting of the hologram authenticity label used on Microsoft's MS-DOS 5 packaging, encouraging the company to adopt a succession of increasingly sophisticated security holograms over the following years, turning buyers into forensic scientists.[65] The International Hologram Manufacturers' Association (IHMA), founded in 1992 to police the industry, waged a running battle against counterfeit security holograms while simultaneously marketing their value to businesses. As one means of stemming the rising tide of counterfeits, the organization instituted a register of commercial holograms.[66]

10.4 THE DIGITAL TRANSITION

As traced throughout this book, twentieth-century visual wonders developed alongside the equally seductive marvels of electronics. Hobbyists' exploration of photography was paralleled by radio-building; amateur telescope-making, unsettling graphic arts and early television were experienced by the same audiences; and hologram exhibitions peaked as personal computers entered popular consciousness.

These co-existing technologies merged together for Cold War engineers, who had first developed holograms as an alternative to slow and temperamental electronic computers. Image processing began not with software, but with holograms (Section 4.4). On the other hand, optical engineers at IBM and in the Soviet Union during the 1960s and seventies found that they could create holograms by computer calculations that modelled the interference of light waves.[67]

[64] Bauer, R., 'Holographie – Eine kommerziell erfolgreiche aber als Kunstform gescheiterte Technologie?', in: R. Bauer, J. Williams and W. Weber (eds.), *Technik zwischen Artes und Arts [Technology between artes and arts]* (Münster: Waxman, 2008), pp 133–47.

[65] A 2015 webpage instructs purchasers of software on DVD to examine 'holographic security features', which included 'vibrant color and 3D effects as the disc is tilted', an 'inner mirror band hologram' on the data side and 'additional optical security features can be seen from both sides of the disc, near the inner hole of the disc and along the outer edge. The features are embedded within the disc; they are not on a sticker'. These optical assurances were to be further checked by a Certificate of Authenticity and a 25-digit Product Key; see http://www.microsoft.com/en-gb/howtotell/Software.aspx#Packaging, accessed 23 Mar 2015.

[66] Lancaster, I. M. and L. T. Kontnik, 'International hologram manufacturers Association (IHMA): hologram counterfeiting and the hologram image register', presented at *Fifth International Symposium on Display Holography*, Lake Forest, IL, 1994; Lancaster, I. M., 'The future for security applications of optical holography', *Proceedings of SPIE* 2577 (1995): 71–6.

[67] Brown, B. R. and A. W. Lohmann, 'Computer-generated binary holograms', *IBM Journal of Research and Development* 13 (1969): 160–8; Yaroslavsky, L. P. and N. S. Merzlyakov, *Methods of Digital Holography* (New York: Consultants Bureau, 1980); Yaroslavsky, L. P., 'Digital holography: 30 years later', *Proceedings of SPIE* 4659 (2002): 1–11; Bove Jr., V. M., 'Display holography's digital second act', *Proceedings of the IEEE* 100 (2012): 918–27.

Such methods were impractical even into the 1990s, as computer speeds and plotting precision could not match traditional optical methods of producing holograms. A new approach was pursued from the late 1980s, however. Instead of creating a hologram holistically by scattering light from a solid object, a three-dimensional image can instead be built up from individual spots of light. Such spots can be produced by a hologram that consists of separate pixels, each of which diffracts light of a particular colour and direction to the viewer's eye. To create such an *elemental* or *dot-matrix* hologram, a computer is used to synthesize the required pixels and to direct an optical or electron-beam printer to record them one by one onto the hologram surface. The resulting digital hologram can be viewed or replicated in the same way as conventionally-produced optical holograms.[68]

The development of the new digital technology was particularly rapid in the Far East, and soon became a global industry of hologram design services and providers of proprietary hologram printing machines that built on the existing network of embossed hologram firms.[69] The first commercial products appeared in 1991. China's burgeoning economy during the 1990s soon made it the world centre for digitally-mastered security holograms and their counterfeits.

Suppliers vied to optimize the sophistication and speed of creation of their hologram creations. The characteristics of digital holograms rapidly diverged from their optical counterparts, becoming imbued with bright and intricate designs of arbitrary complexity. Images could be generated from computer-simulated imagery, and so provided an ideal link between the increasingly digital world of the 2000s and real-world enjoyment (Figure 10.11).

As with the market for conventionally-made embossed holograms, their digital cousins boosted the authentication industry. But image complexity could protect identity documents and designer products only briefly. Hologram generation and printing equipment could increasingly be purchased as turn-key systems by illegitimate as well as legitimate firms. Counterfeiters proved able to create fraudulent holograms in days, if not hours.

Security and packaging holograms invaded contemporary culture from the 1980s. But this was a top-down phenomenon. The social and economic need for these optical devices has been promoted by corporate and governmental interests, and protected by them against forgeries and interloper technologies. It was a fragile, dynamic stability.

Technical limitations also remained. Despite prodigious efforts to create optimized computer hardware, as of 2015 digital holograms calculated in real-time do not yet have the resolution or impact of pre-recorded holograms.[70] And, while digital holograms could be made brighter and viewable at wider angles than their optically-generated

[68] Newswanger, C. D., 'Dot matrix technique for generating diffraction grating patterns', presented at *Fifth International Symposium on Display Holography*, Lake Forest, IL, 1994; 'Production equipment: dot matrix hologram machines', *Holography Marketplace* 8 (1999): 105–10.

[69] See, for example, Li, Y., T. Wang, S. Yang, S. Zhang, S. Fan and H. Wen, 'Theoretical and experimental study of dot matrix hologram', *Proceedings of SPIE* 3559 (1998): 121–9.

[70] For an optimistic early forecast of computer-generated holographic environments, see Lecht, C. P., 'A man's hologram is his castle: emergence of artificial experience', *Computerworld*, 17 (14), 4 Apr 1983: 47–50.

Fig. 10.11 Computer-generated surface-relief holograms on adhesive foil (Manufactured by Security Imaging Ltd, 2002; S. Johnston photo).

counterparts, they still required more careful lighting and viewing than photographs did. Like the Lippmann photograph a century earlier, the appearance of colour in most digital hologram images depended sensitively on angle, making faithful reproduction impossible. Thus computing capabilities have continued to improve the sophistication of synthesized digital holograms, but their optical limitations, recognized for a half century, have proven more difficult to overcome. Yet even if commercial holographic products were a short-lived explosion, there was enduring cultural fallout: holograms could continue to expand in popular imagination.

11

Channelling Dreams

The magazine's name really owes its origins to two parts. The Tales *part is a tribute to the old 1930s pulp science fiction magazines* Weird Science Tales, Amazing Tales, *etc. The* Hologram *part actually owes its origins to the cyberpunk author William Gibson's seminal SF short story of the 1980s, 'Fragments of a hologram rose'. I blended the two together, and came up with* Hologram Tales. *In the words of JRR Tolkien, when asked why he called his main race "hobbits" . . . 'I just thought it sounded nice'.*[1]

Somewhere on the internet, 2002

From the explosion of popular attention that began in the 1960s, holograms developed both real and imaginary parts. Forecasts of their development, so central to engineering cultures and corporate promotion, equally inspired speculative fantasy and popular science fiction. Instilled with imagined attributes that reflected cultural desires and expectations, holograms of the future developed a virtual life. They carried on the modernist tradition: in the worlds of fiction expressed in new media, they represented unbridled technological progress that seemed inevitable. But holograms also evoked other associations: older technologies, contemporary fashions, rumours and myth. Although genuine holograms are less commonly viewed today, their symbolic identity has been grafted onto an expanding range of cultural products and anticipations.

Tracing this cultural evolution takes us further from the scientific roots of holograms. Instead of representing an inexorable expansion of knowledge and application, holograms began to accrete competing popular meanings. The anthropologist Claude Lévi-Strauss dubbed this combining of cultural notions into new understandings *bricolage*. In popular usage, the French term is usually translated as 'practical innovation' or 'do-it-yourself': a technical activity associated with non-professionals—the very activity that Albert Ingalls and C. L. Stong at *Scientific American* vaunted as technical 'tinkering'. As employed by Lévi-Strauss, *bricolage* was more pointedly an analogy of how primitive minds ('la pensée sauvage') construct myths. Behind his use of the term lies the idea that popular thought is inferior to rational understandings.[2]

[1] Hunt, S. to S. F. Johnston, email, 13 Dec 2002, SFJ collection.

[2] Lévi-Strauss, C., *La Pensée Sauvage* (Paris: Librairie Plon, 1962); Johnson, C., '*Bricoleur* and *bricolage*: from metaphor to universal concept', *Paragraph* 35 (2012): 355–72.

A related and more even-handed concept is the *meme*. As introduced by biologist Richard Dawkins, the meme was a metaphor to suggest that cultural beliefs and practices spread much as biological genes do. Memes are 'idea chunks': fragments of concepts that are incomplete in themselves but capable of combining into robust beliefs that thrive in particular cultural environments. Like biological genes these durable notions are transmitted by being copied by their hosts; popular ideas are passed on and so propagate through culture. The evolutionary metaphor suggests some important effects. This mutation of ideas can generate unexpected syntheses; the resulting cultural notions may overwhelm those produced by more rational methods; and—just as outbreaks of infectious disease have been liberated by jet travel—memes are transmitted with great efficiency by the internet.[3]

Methods of disseminating memes early in the century had relied on publications and their publishers. As described in Chapter 7, for example, influential organizations such as Science Service and *Scientific American* could promote technical pastimes and shape a generation. Cultural fashions can be, more often than recognized, engineered by specific actors. This hierarchical dimension is even less commonly identified, but no less potent, for popular notions spread via the internet. As with other forms of social network, the attitudes and beliefs of influential individuals blogging or tweeting may be emulated by peers sharing their interests. High esteem in the domain of online gaming or fashion can reinforce even abstract convictions such as technological optimism or obeisance to conspiracy theories.[4]

A more anthropologically grounded perspective is provided by the field of cultural ecology, which focuses on social organization and contexts as key factors affecting how themes evolve in culture. Popular ideas flourish or starve within different environments, co-evolving with the social groups who nurture them. Like the concepts of *bricolage* and the meme, cultural ecology has been explored not only within anthropology but also in literary studies and related fields. This approach recognizes a number of the features relevant to holograms, notably how they have been conceived and used by distinct social groups, and how they link to established social practices of the past.[5]

11.1 GALAXIES AND HOLOGRAMS FAR, FAR AWAY

While public engagement with holograms was still rising during the late 1970s and early 1980s, a second and distinctive understanding was emerging. Its source was science

[3] Dawkins, R., *The Selfish Gene* (Oxford: Oxford University Press, 1976); Lynch, A., *Thought Contagion: How Belief Spreads Through Society* (New York: BasicBooks, 1996); Blackmore, S., *The Meme Machine* (Oxford: Oxford University Press, 1999); Shifman, L., *Memes in Digital Culture* (Cambridge, MA: MIT Press, 2013). Dawkins has suggested the internet memes differ from his original concept because they can be consciously created and seeded, unlike earlier forms that were more analogous to random selection and mutation.

[4] See, for example, Wasserman, S. and K. Faust, *Social Network Analysis: Methods and Applications* (Cambridge: Cambridge University Press, 1994).

[5] For an introduction to the field, see Duncan, J., *A Companion to Cultural Ecology* (New York: Wiley-Blackwell, 2007).

fiction, and the initial ideas were absorbed from the contemporary forecasts of optimistic technologists.

As early as 1968, John Brunner's novel *Stand on Zanzibar* described a 'holocam' for recording three-dimensional images, 'lit by LazeeLaser monochrome lamps'.[6] The first sci-fi story to focus a plot on holograms appeared four years later: Ed Bryant's 'The poet in the hologram in the middle of prime time' cast it as a form of three-dimensional interactive television, 'holovision'. The imagined technology is a recognizable echo of the forecasts disseminated by Conductron Corporation six years earlier and, as the author suggests, it could readily be added to consumer dreams by promoters:

> 'Unicom's new line of holovision sets for 2020!'
>
> 'Oh dear,' said the wife. 'We already have a holovision set.' There was regret in her voice at having to disappoint the salesman who looked so much like her favourite uncle.
>
> 'Not like this one, you don't!' The salesman pivoted and dramatically indicated a shining black box on a crystal dais.
>
> 'Friends, you undoubtedly have an old-style holovision—the kind that only gives you three-dimensional pictures and stereo sound.'
>
> 'Of course,' said the husband, puzzled. 'It's the best set on the market'.
>
> 'Not any more! Not now that UniCom has added a whole new dimension to holograms!'
>
> The prospective customers appeared properly astonished and intrigued. 'A new dimension?' they asked in concert.
>
> 'Brand new! It's now possible for you—' he pointed to the woman. 'And you—' he gestured to the man. 'To actually participate, to star in your own favorite holovision shows, right in the comfort and convenience of your own home.'
>
> The couple looked struck by wonder.
>
> 'Imagine—' said the woman.[7]

By the late 1970s and during the proliferating public exhibitions, however, holograms began to embed in popular imagination. Set in the indefinite future, William Gibson's meticulously detailed fictional worlds incorporated holograms in roles that had become familiar from technologists' predictions but also in new forms.

Gibson's short story 'Fragments of a hologram rose' imagined holograms as small laminated plastic postcards much like those that were beginning to appear in the shops of 1977:

> Parker lies in darkness, recalling the thousand fragments of the hologram rose. A hologram has this quality: recovered and illuminated, each fragment will reveal the whole image of the rose.[8]

[6] Brunner, J., *Stand on Zanzibar* (London: Doubleday, 1968), p. 33.

[7] Bryant, E., 'The poet in the hologram in the middle of prime time', in: H. Harrison (ed.), *Nova 2: Original Science Fiction Stories* (London: Sphere Books, 1972), pp 116–32; quotation p. 124.

[8] Gibson, W., 'Fragments of a hologram rose', in: W. Gibson (ed.), *Burning Chrome* (New York: Ace Books, 1987), pp 31–9, first published in the short-lived science fiction magazine *UnEarth 3* in 1977.

In subsequent stories, though, Gibson invests holograms with qualities that have been familiar in science fiction ever since. Inventing holographic novelties and just as rapidly jumping beyond them, he imagines 'Disney technicians . . . employed to weave animated holograms of Egyptian hieroglyphs into the fabric of your jeans' ('The Gernsback continuum', originally published 1981), 'peeling chrome over plastic, blurry holograms that gave you a headache if you tried to make them out' and 'a defective hologram in the window, METRO HOLOGRAFIX, over a display of dead flies wearing fur coats of gray dust' ('Burning Chrome', originally published 1982).[9]

Gibson's cyberpunk world—a run-down and cynically-peopled urban environment that he dubbed 'the sprawl'—provided both a contemporary judgement and futuristic prediction for holograms. These consumer items and advertising media were tawdry and twee. He predicted that their powers to entertain the public would wither:

> Holography went too, and the block-wide Fuller domes that had been the holo temples of Parker's childhood became multilevel supermarkets, or housed dusty amusement arcades where you still might find the old consoles.[10]

Even before the real-world holography industry had found its feet, then, Gibson's fiction forecast an uninspiring future. His fiction highlighted this perspective that countered the modernist visions of the 1930s. In fact one story, 'The Gernsback continuum', contrasted the faded dreams of Hugo Gernsback's technology magazines (Section 7.2) with the gritty reality of the present.[11]

Other science fiction writers took up and expanded the tropes as incidental but sophisticated elements of their imagined futures. For John Varley, 'holomist' and 'ghostsmoke' were forms of advertising that hounded prospective customers, playing 'breath-catching tricks with perspective'. For Philip K. Dick, a 'holo-cube' was a computer-generated immersive environment that could be called up, entered or collapsed at will. Arthur C. Clark conceived a 'holopad', a medical monitor with three-dimensional display. And adopting the tone of William Gibson, a Tom Maddox novel described future holograms as typical examples of corporate development for jaded consumers: 'people a foot high chattered . . . under the light of a Blaupunkt holostage'. Writers rapidly adopted the prefix *holo-* for further imagined inventions, consolidating these concepts in readers' minds.[12]

For more optimistic audiences, science fiction could portray holograms as elements of a visually exciting technological world; cinema, television and videogames could

[9] Gibson, W., *Burning Chrome* (New York: Arbor House, 1986), quotations pp 23, 166 and 195, respectively.

[10] Gibson, W., 'Fragments of a hologram rose', in: W. Gibson (ed.),*Burning Chrome* (New York: Ace Books, 1987), pp 31–9; quotation p. 34.

[11] Gibson, W., 'The Gernsback continuum', in: W. Gibson (ed.), *Burning Chrome* (New York: Arbor House, 1986), pp 24–36.

[12] Varley, J., *The Ophiuchi Hotline* (New York: Dial Press, 1977), p. 96; Dick, P. K., *A Scanner Darkly* (New York: Doubleday, 1977), p. 179; Clark, A. C., *The Fountains of Paradise* (New York: Ballantine, 1978), p. 33; Maddox, T., *Snake-Eyes* (New York: Omni Publications, 1986), p. 45. *Holo* as an abbreviation for holographically-displayed information appears to have had its first fictional reference in Niven, L., *Ringworld* (New York: Ballantine, 1970), p. 31.

illustrate what the fiction writers had described. Their visual properties were forecast compellingly with the release of *Star Wars* in 1977. As revealed by early drafts of George Lucas's screenplay two years earlier, the model resembled a development of the multiplex hologram:

> He sits up and sees a twelve-inch three-dimensional hologram of Leia Organa, the Rebel senator, being projected from the face of little Artoo. The image is a rainbow of colors as it flickers and jiggles in the dimly lit garage. Luke's mouth hangs open in awe.[13]

The link to contemporary Multiplex holograms is notable, but the scene added a more compelling technical feature: the notion of a *projected* hologram rapidly became a science fiction trope. It had not been part of the forecasts by engineers of the 1960s and seventies: they had acknowledged that holographic schemes would be limited by the screen, or hologram itself, both framing and constraining the view. But the cinematic simulation, combining a cone of light from the imagined projector, a juddering blue-cast image evocative of a badly-tuned black-and-white television set and—most impressive of all—the creation of this spectre in empty space, made the vision believable for theatre audiences. These imagined properties created a link between holograms and electronic media for the viewing public. The *Star Wars* sequels imagined further uses: holograms as static displays, as control panel elements and as improved playback devices that could display recorded holograms as magnified or reduced imagery. Conforming to the descriptions in William Gibson's imagined worlds, cinematic holograms adopted the convention of being stuttering, run-down and imperfect. Ironically, this motif of the flawed visual image was conducive to the public reception of faux 'entertainment holograms' of the 2010s (see Appendix, Section A.2), which looked recognizably artificial next to their live counterparts.

Star Wars, sequels and prequels were followed by a wave of fantasy holographic characters. The British comedy television series *Red Dwarf* (pilot written in 1983, series broadcast in Britain 1988–94), set on board a spaceship of the distant future, allied the attributes of fictional holograms to new technologies becoming familiar to viewers. One of the characters of the series, Lister, is a hologram calculated in real time by the spaceship's computer. His three-dimensional image is painted by a 'light-bee', the image melding seamlessly with the AI voice and personality. Tapping into viewers' expectations of progress, the technology even improves: in early episodes the character—described as being generated by 'Soft Light'—cannot be touched; later episodes reveal him to have been rewired with a 'Hard Light drive' to generate a sense of solidity. The stories echoed viewers' experiences with modern technologies.[14] The series overlapped the boom in home

[13] Lucas, G., 'Star Wars script (revised fourth draft)', 15 Jan 1976. The notion of a projected 3-D character was much older, though: the earliest may be the 1928 account in *Weird Tales* (Indianapolis: Popular Fiction Publishing) of the *telestereo*, a 'glass disk, inset in the room's floor' upon which appeared 'the image of a man in the blue and white robe of the Supreme Council, a lifesize and moving and stereoscopically perfect image, flashed across the void of space to my apparatus by means of etheric vibrations' (reprinted in Hamilton, E., *Crashing Suns* (New York: Ace, 1965), quotation p. 30).

[14] Naylor, G., *Red Dwarf*, (BBC Two, 1988–99).

computing which had been triggered by the explosion in affordable hobbyist 'microcomputers' such as the Radio Shack TRS80 (1977), Apple II (1977) and Commodore 64 (1982) in America and the Sinclair ZX80 (1980) and BBC Micro (1981) in Britain. Over the following decade, a generation of hobby programmers wrote their own games, appropriating the fantasy technologies and storylines. Thus science fiction consumers and amateur technical enthusiasts—both significant social alignments of the century—spawned the creation of the futuristic hologram from a jostling collection of cultural tropes.

The rapidly extending specifications of fictional holograms mirrored not just the quickly evolving personal computer scene, but also contemporary holographic research: Stephen Benton's group at the MIT Media Lab was then exploring the feasibility of haptic interfaces, giving holographic images the illusion of substance. The cultural notion of the hologram, in both fiction and engineering aspirations, was generalizing from a mere optical effect to one imbued with mechanical and electronic underpinnings.[15]

Similar dreams were a part of the second *Star Trek* television series *The Next Generation* (broadcast 1987–94) with its stories of a 'holodeck', or holographic visualization room, in which an entire environment can be calculated and displayed by computer (and, indeed, interacted with by the human characters). The Star Trek television storytelling diverged increasingly from the prospects for real-world holograms, much to the chagrin of practicing holographers, who found themselves increasingly challenged by the preconceptions and disappointments of the general public. As in *Red Dwarf*, the fictional technology was gradually augmented through the series by new capabilities. The holodeck environment was now a combination of both optical and physical creations, some described as being 'replicated' (that is, material objects manufactured from individual atoms) and animated with 'weak tractor beams' and 'shaped force fields'. In the succeeding series, *Deep Space Nine* (1993–99), the technology becomes commercialized: 'holosuites', or holodecks of various capacities, are rented out for private use. And in the *Voyager* series (1995–2001) these technologies are combined with artificial intelligence (AI) to implement an Emergency Medical Hologram, or virtual doctor. Such sentient computer programs embodied in a computer-generated hologram became a commonplace idea—even a convention—in subsequent fiction.[16]

For a younger generation, the animated television series *Jem and the Holograms* (1985–88) imagined holograms as a technology of self-empowerment. The plots centre around

[15] See, for example, Plesniak, W. J. and M. A. Klug, 'Tangible holography: adding synthetic touch to 3D display', *Proceedings of SPIE* 3011 (1997): 53–60; Pappu, R. and W. Plesniak, 'Haptic interaction with holographic video images', *Proceedings of SPIE* 3293 (1998): 38–45.

[16] Paramount Domestic Television, *Star Trek: The Next Generation* (CBS, 1987–94); Paramount Domestic Television, *Star Trek: Deep Space Nine* (CBS, 1993–99); Paramount Domestic Television, *Star Trek: Voyager* (CBS, 1995–2001); see also Picardo, R., *Star Trek Voyager: The Hologram's Handbook* (Pocket Books, 2002). More recently, holographic characters have featured in contemporary settings, too. In an episode of *Perception*, the abilities of modern-day artificial-intelligence and hologram researcher are embodied in a holographic double; when the professor is murdered, his doppelganger helps to solve the crime (ABC Studios, 'Eternity', *Perception*, series 3, episode 5 (TNT, 2014)).

a holographic computer and projection device, *Synergy*, having characteristics akin to *Red Dwarf*'s light bee and *Star Trek*'s holodeck. With it, the principal character, Jem, is able to transform her local environment and cloak her identity. The result is alternately an 'entertainment synthesizer' and shape-shifting device for adventures. It is worth noting that the term 'shape-shifting', which sprouted in popular fiction from the 1970s, is an ancient trope traditionally linked to magic and sorcery. Thus fantasy holograms, like their real-world counterparts, grafted folklore, conjuring and modernity.[17]

Video games from the 1990s began to transmit such visions to increasingly wide audiences. Some games extended real-world engineering claims, while others rehearsed the metaphors that had been introduced in science fiction cinema and television. Holographic head up displays (HUDs), for example, became a common feature of most first-person shooter videogames: weapons, vehicles or tools would be selected from a floating palette of choices, mimicking a visual effect that can be produced in the cockpit of a military aircraft or a more mundane teleprompter.

Other imagined hologram technologies are similarly gentle extensions of the state-of-the-art. The PlayStation series *Ratchet & Clank* (2002–) incorporated 'holo-plans', three-dimensional blueprints of weapons. The imagined holo-maps are more compact than three-dimensional models and can be assembled to reveal complex designs—similar in usage to real-world computer-generated hologram tiles developed as tactical maps for military and urban planning (Figure 11.1).[18]

The same game series also incorporates holograms of the *Star Wars* style, replete with the flickering colours and tinny echoing sound of a malfunctioning television or video-recorder of the 1970s. Thus, the retro qualities of the Princess Leia image—so familiar for the baby boom generation—were appropriated for their children.

Interestingly, holograms were never more than incidental plot elements in videogame series. As in William Gibson's fiction, the *Ratchet & Clank* holograms merely underline a futuristic aesthetic, acting as window-dressing rather than key functional features. Other videogames, however, added new attributes to holograms. The most influential was the notion of the hologram as a synthetic mirage. Camouflage, especially through optical confusion, had become an important technological ploy during the First World War, and game holograms extended this heritage.[19]

[17] Hasbro, *Jem and the Holograms* (Claster Television, 1995–2001). On the usage of the term 'shape shifting', see Google ngrams, https://books.google.com/ngrams/graph?content=shape+shifting&year_start=1900&year_end=2008&corpus=0&smoothing=3&share=&direct_url=t1%3B%2Cshape%20shifting%3B%2Cc0, retrieved 19 Jun 2014.

[18] Insomniac Games, *Ratchet & Clank* videogame (Sony, 2002).

[19] *Camoufleurs* ('disguise' artists) comprised the original French camouflage corps in 1914, an initiative soon paralleled by other combatants. These early optical experiments were not meant to hide or cloak targets. Clothing, equipment shrouds, trucks and even battleships were painted in contrasting and often dazzling patterns that sought to confuse the recognition of edges, shapes and motion—another novel application of visual surprise. For examples of such 'disruptive' camouflage, see Behrens, R. R. (ed.), *Ship Shape, a Dazzle Camouflage Sourcebook: An Anthology of Writings About Ship Camouflage During World War I* (London: Bobolink, 2012). Such optical trickery was founded on Victorian research on optical amusements. Thus zoetropes, military camouflage and holograms shared common roots and cultural appeal.

Fig. 11.1 Life imitating art. Horizontal full-colour reflection hologram of downtown Seattle generated from computer data, lit from above by a white light source ().

It is worth noting that the popular idea of holograms as mirage, like holographic television, was not a popular invention, but was introduced by engineers who recognized the technical challenges. During the 1960s, a handful of proponents explored the concept of holograms as mirages. Kip Siegel of Conductron Corporation had first suggested holograms as warning signs, and the idea was rapidly taken up by other entrepreneurs. A 1969 *Popular Science* article touted the potential as reality (Figure 11.2):

> **Want to stop traffic? Holosigns do, by projecting images in midair, without a screen!**
> Holosigns are glareless. They also penetrate rain and snow when conventional signs are blotted out . . . The device will permit signs to appear where none were possible before. Airports could use a succession of Holosigns to simplify aircraft landings . . . Two different movies could be shown in one theatre, by beaming an adult picture at one section and a children's picture at another.

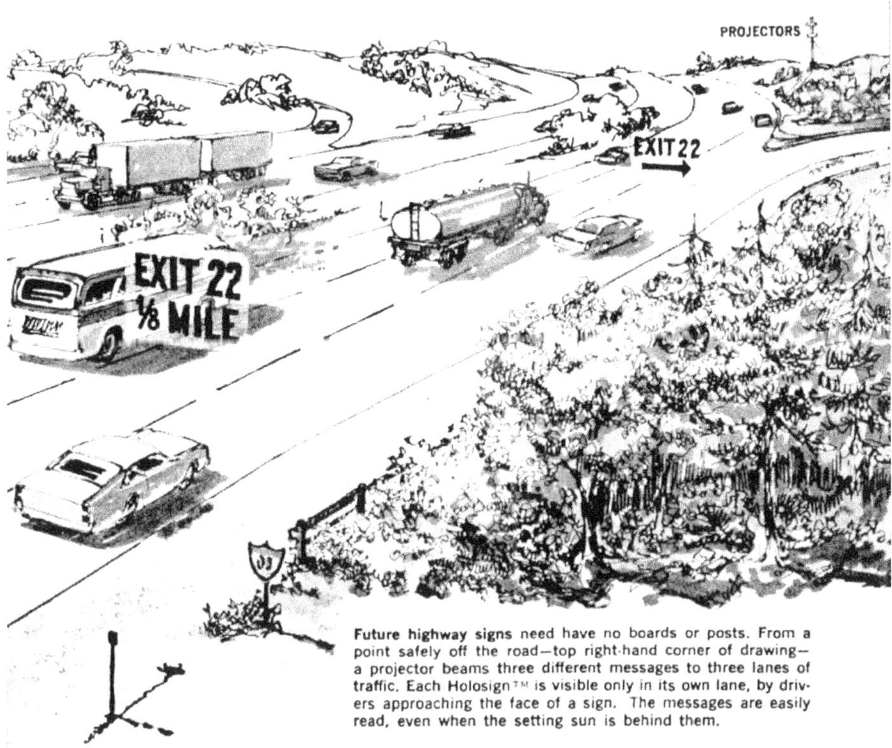

Fig. 11.2 Imagining holographic mirages (Arctander, E. H., 'Signs you'll see appearing out of thin air', *Popular Science*, 195 (2), Aug 1969: 81–2, 193, illustration p. 81 by Dana Rasmussen).

The claims deployed by its inventor, Harry Forster of Holograph Corporation, triggered readers' attractions to hidden technologies and their powers:

> 'Holosigns can be described in terms of holography, but we have modified the process', Forster says mysteriously . . . He prefers not to talk about what's inside the box . . . 'We're working in a never-never land because people haven't been here before'.[20]

Would-be customers took the bait:

> A fellow from the public roads department came wanting us to provide expertise on a scheme he had for putting up holograms on the expressways. He wanted to have them project a 3-D brick wall for people wanting to drive up the wrong exit ramps the wrong way. I tried to convince him that it would take a hologram the size of a billboard and a very bright light source to make that really effective. I said that holograms of that size had not been made, but that

[20] Arctander, E. H., 'Signs you'll see appearing out of thin air', *Popular Science*, 195 (2), Aug 1969: 81–2, 193; quotations pp 81–2. A related patent was later granted but—unlike the article claims—the invention did indeed require screens and was not inherently holographic (Forster, H. D., Pat. No. US3844645 A 'Three-dimensional projecting system' (1974), assigned to Holograph Corporation).

> perhaps one could put together a mosaic of them and do it that way. Although I found the scheme very impractical and tried to tell him so, he would have none of it. He had gone out on a limb, so to speak, in his department, and his career there was at stake. They had even already put it out for bids and one company in California (where else?) had submitted a paper saying they were willing to study the question (for a price, of course). The fellow went ahead and convened a panel of experts which included me and a group of physicists from the Goddard Space Flight Center and we all tried as gently as we could to tell him his scheme would not work. He cried. We suggested other things he might think of doing to improve the roads, but he was not able to listen to us.[21]

Such tales of promoter hype and investor naïveté, recounted by a generation of jaded engineers, became contemporary legends in their own right and spawned further mythology. Similar notions have been revived whenever the demands of military advantage or commercial innovation have motivated uncritical sponsorship. A more influential, but equally ephemeral, example is a single *Washington Post* article that reported a discussion paper presented at an Air Force conference. The outline of the 1994 presentation imagined the technologies required thirty years hence to make an 'airborne holographic projector' and its potential use for psychological warfare against a naïve enemy.[22]

Video games dedicated to battlefield simulations became a platform for such ideas. In *Command and Conquer: Red Alert 2* (2000), 'mirage tanks' project holograms to masquerade as trees. Similarly, in *Halo* (2001), holograms create decoys, or cloak facilities.[23] Gameplay holograms mix and match properties promiscuously. In the videogame Crysis 2 (2011) a 'holographic decoy' is a weapon attachment that can emit or spawn a replica soldier to attract gunfire. An apparently animate and autonomous mirage, its characteristics are similar to the Emergency Medical Hologram of *Star Trek: Voyager*. And, in an even more dynamic role, the game *Duke Nukem 3D* (1996) includes the 'HoloDuke' device which projects a holographic mirage that can battle opponents. Its solidity is reminiscent of later versions of *Red Dwarf*'s holographic character.[24]

As such examples suggest, videogame holograms are not formulaic. From game to game, their attributes shift fluidly and draw upon a broad range of cultural understandings. Some—like the gently humorous *Ratchet & Clank* for younger players and more sophisticated cultural parodies in *Duke Nukem 3D*—play on, and poke fun at, shared understandings, thereby strengthening gamers' sense of community. Multi-player games accelerate the creation and evolution of this shared identity.

[21] Funkhouser, A. to SFJ, email, 9 Apr 2003, SFJ collection, quoted; A similar anecdote was related by Cochran, G. D. to SFJ, interview, 6 and 8 Sep 2003, Ann Arbor, MI, SFJ collection.

[22] Arkin, W. M., 'When seeing and hearing isn't believing', *Washington Post*, 1 Feb 1999, http://www.washingtonpost.com/wp-srv/national/dotmil/arkin020199.htm, accessed 11 Dec 2014.

[23] Westwood Pacific, *Command and Conquer: Red Alert 2* videogame (Microsoft, 2000); Bungie, *Halo* videogame (Microsoft, 2000). The notion of optically masking an object also originated in science fiction at least thirty years earlier. The original *Star Trek* series depicted a Romulan vessel made invisible by an unknown technology in the 1966 episode 'Balance of Terror' and subsequently identified as a 'cloaking device' in the 1968 episode 'The Enterprise Incident'.

[24] Crytek, *Crysis 2* videogame (Electronic Arts, 2011); 3D Realms, *Duke Nukem 3D* videogame (3D Realms, 1996).

As for an earlier generation of gullible investors and consumers, some gamers will not distinguish technological dreams from parody. When experienced alongside the more familiar technologies featured in such storylines, these immersive fantasies could become so compelling that they escape the gaming world to blur with reality. The rise of such technological myths has been further reinforced by the internet. Where multi-player games allowed real-time cooperative gameplay, internet blogs, discussion groups and other social media encouraged lengthier dialogue. Holograms have been liberated by online fantasy.

The arrival of the internet allowed unmediated—and unverifiable—claims to reach receptive audiences via social media. Flourishing subcultures of gamers could invest their simulated worlds with complex perspectives drawn from real life; on the other hand, less worldly fans increasingly confused the imaginable with the feasible, and even the inevitable.

Owing to the enculturation provided by videogames and the wildfire proliferation and mutation of online information, forecasting the future has become a malleable and democratic art. Popular myths that the American military is developing, or has even perfected, holographic 'projections' to frighten or confuse the enemy became an internet meme, having originated from a combination of institutional musing, science fiction films and video-gaming.[25] From the late 1990s, online forums elaborated the tale into an advanced covert project aimed at deceiving foreign or domestic audiences ('Project Blue Beam', ostensibly under the aegis either of NASA, the Defense Advanced Research Project Agency (DARPA) or United Nations). Holograms satisfy the distinct appeals of hidden science, magical portrayals and conspiracy theories. Thus optimism and credulity, in equal measure, have dogged the imagined holograms of the twenty-first century.[26]

From the revelation of the hologram in the 1960s, then, there has been an increasingly obvious bifurcation of technologists' and science fiction writers' conceptions. They extended the ontology of popular fantasies. Holograms became a staple of science fiction plots alongside robots, time travel and wormholes. But, in the process, holograms were translated into a more compelling technology for contemporary audiences. The future would include more immersive and interactive experiences; fictional holograms would integrate computing and intelligence, and appeal to additional senses. The holograms of popular fantasy were forecasts allied with particular worldviews, like the stories in which they were embedded. The hologram might represent mythical magic, an imminent new channel

[25] As popularized by Canadian journalist Serge Monast from 1994, the project was said to involve displaying 'holographic' images of sacred figures to compel the obedience of an unsophisticated population. The notion has notable similarities, however, to a film treatment for a proposed *Star Trek* movie, *The God Thing*, discussed publicly around the same time (Engel, J., *Gene Roddenberry: The Myth and the Man Behind Star Trek* (New York: Hyperion, 1994)).

[26] Since 1996, the term 'holographic projection' has trebled in usage in English language publications (Google ngrams, https://books.google.com/ngrams/graph?content=holographic+projection, retrieved 19 Apr 2015).

for mass consumption, the decline of modernist dreams or the secretive tool of a militaristic state but, in one guise or another, it came to inhabit most of our imagined futures.

11.2 SPECTACLE, AESTHETICS OR UBIQUITY?

My eyes gradually accommodated to the dim interior of the former warehouse to reveal airy galleries hung with holograms. Most of the works on display dated from the 1980s and encapsulated the excitement surrounding the creation of those colourful and often dynamic images. At the centre of one gallery, a lonely transmission hologram, lit awkwardly by an inadequate laser, produced a distorted image in a shrouded enclosure. In the lecture room a large audience was gathering, attracted by Facebook *and* Twitter *feeds rather than the radio advertisements and posters that had announced hologram exhibitions in the past century. My talk was being streamed to a student online channel in London. As I spoke about the old environments and their artistic creators, the twenty-something audience seemed more interested in the near future than the past. During the question period, one asked, 'Are the holograms on CNN News typical of what's coming?'*

Glasgow, 2012

Stretched by popular anticipations, promoters have always lamented the difficulty of explaining holograms to wider publics and bemoaned the misidentification of other technologies as holograms. A half-century after that futile dialogue began, the term *hologram* in pop culture has evolved from optical tricks to security patches to virtual entertainment in live settings.[27]

But, like the challenges to the Académie Française to fully regulate the official French language, the borders of technical jargon can never be policed effectively. Current usage is imprecise and all-encompassing: the label may be applied to any ephemeral imagery associated with three-dimensionality, hanging in space and having an aura of mystery or spectacle. The term labels a cultural construct that has much greater potency than the scientific product itself.[28]

Even technical terms have limited shelf-life. As concepts shift, specialist language follows, mutating the meaning of jargon that had initially defined clear categories or entities. In the USA, a 'xerox' became the generic term for a photocopy after Xerox Corporation founded and dominated the market for electrostatic copiers during the 1960s; for the same reason, 'hoover' entered common usage (again as both noun and verb) in the UK as a synonym for vacuum cleaner.

[27] Surveying my undergraduate students over some fifteen years has revealed waves of meanings associated with particular age groups. For those born during the 1960s, holograms were most often identified as art forms and portraits; for students born during the 1970s to early 1990s, either security patches or science fiction were the dominant evocations; and for those born post-1995, holograms were most frequently linked to virtual entertainment in traditionally live settings, such as the CNN interviews employing telepresence, and recreations of deceased pop performers, both touted as 'holograms'.

[28] The Hellenic Institute of Holography has abandoned the term *hologram* in its publicity in favour of a neologism, *OptoClone*™.

One reason that the term *hologram* proved to be so generic and unstable is that holographers themselves devised and promoted a bewildering variety of minor variants in rapid succession—indeed, the commonalities continue to exercise patent attorneys and to frustrate attendees at technical conferences. Some of the forms relied on differences of manufacture scarcely identifiable by non-specialists. Others evinced subtle differences of optical properties that allowed them to masquerade as each other.

During the late 1960s, the notion of 'white-light holograms' circulated between development laboratories. For the viewer, their most obvious attribute was either appearing in gradations of brightness (e.g. like a nearly monochrome three-dimensional image) and often having the image straddling the hologram surface, or being lit (reconstructed) by a white-light source such as a light bulb. Distinct technical processes could produce similar visual effects. For example, the term was first used by engineers to describe an image focused near the hologram plane.[29] Alternatively, it could label reflection holograms lit by a white-light source (known as Lippmann, Lippmann–Bragg or Denisyuk holograms).[30] And, over the following decade, the label was increasingly appropriated by investigators following the work of Stephen Benton at Polaroid, who used the term 'white light transmission' (WLT) hologram to describe his recording technique that relied on a strip aperture to record a holographic image that could be viewed with a light bulb in a spectrum of rainbow colours.[31]

To complicate matters further, the various properties could be combined. A hologram might be created on a film or aluminium foil (or even chocolate), transmitting or reflecting an image that could be rainbow-like, plate-straddling or animated. The nascent jargon was imprecise and clumsily re-used. Such distinctions were confusing for optical scientists, corporate public relations officers, patent examiners and non-specialists alike.

From the 1990s, the prefix 'holo-' was used promiscuously not just by science fiction writers, but also by a new wave of commercial products. Consumer displays that showed depth without special glasses were trialled and, by the turn of the twenty-first century, these substitutions had largely edged out real-world holograms in popular consciousness. For their creators, this misidentification is frustrating but seemingly irreversible. I have used the term 'faux hologram' to describe such masquerades.

[29] Stroke, G. W., 'White-light reconstruction of holographic images using transmission holograms recorded with conventionally-focused images and "in-line" background', *Physics Letters* 23 (1966): 325–7; Kakichashvili Sh, D., 'Reconstruction of focused hologram in white light', *Ukrayins'kyi Fizychnyi Zhurnal* 14 (1969): 1862–6.

[30] Denisyuk, Y. N., 'On the problem of a photograph reproducing the full illusion of the reality of the object depicted (in Russian)', *Zhurnal Nauchnoi i Prikladnoi Fotografi i Kinematografi* 11 (1966): 46–56; Stroke, G. W. and A. E. Labeyrie, 'White-light reconstruction of holographic images using the Lippmann–Bragg diffraction effect', *Physics Letters* 20 (1966): 368–70; Denisyuk Yu, N., 'Imaging properties of light intensity waves: the development of the initial Lippmann ideas', *Journal of Optics* 22 (1991): 275–80.

[31] Benton, S. A., 'Rainbow holograms', *Journal of the Optical Society of America* 50 (1969): 1545–6; Benton, S. A., 'Achromatic images from white light transmission holograms', *Journal of the Optical Society of America* 68 (1978): 1441; Benton, S. A., H. S. Mingace, Jr. and W. R. Walter, 'One-step white-light transmission holography', in: P. N. Tamura and T. C. Lee (eds.), *Recent Advances in Holography* (Bellingham, WA: SPIE, 1980), pp 156–61.

The technologies and cultural contexts of holograms (and alternatives that masqueraded as holograms) were interrelated as summarized in Table 11.1. In various guises, the hologram had attributes of earlier visual media—photographs, sculpture, optical toys, cinema, product packaging and graphics—blurring any overarching metaphorical identity. As the table suggests, the evolution of the hologram and its functions is not only a tale of technical advance but also of adaptation to social niches.

In visual terms, it is ironic that laser-lit holograms—the most faithful imagery yet invented—represented the beginning, not the maturity, of the field. The most technical accurate and visually stunning holograms were the Leith–Upatnieks variety, which showed unequalled depth and resolution. These have seldom been exhibited since the early 1970s, however, because of the need for lasers and their consequent ocular safety concerns. Such holograms were subsequently relegated to scientific and engineering laboratories and hobbyists' basements.

The many varieties of 'white light' holograms that followed all compromised image characteristics to obtain a readily available bright light reconstruction. All share one characteristic attribute: a depth of field (zone of sharp imaging) that is diminished when the light source is less monochromatic or not a point source. The imagery of white light holograms always appears sharp in the plane of the hologram, and progressively fuzzier closer or further away. Rainbow holograms have a further limitation, in that they sacrifice vertical parallax (i.e. the ability to peer over foreground elements of the reconstructed scene to see occluded ones). Artisanal and art holographers of the 1970s accepted these constraints in return for the convenience of lighting and the visual surprise of an image that could appear both in front of and behind the hologram plane. Holograms could no longer shock as readily as they had, but they became more affordable and ubiquitous. And, in the hands of artists, rainbow colours could be incorporated pleasingly into abstract art holograms, providing a unique aesthetic function for the medium. This trend was accentuated by the embossed hologram. During the first decade, promoters and some holographers saw them as the means of democratizing the medium and expanding its applications. During the 1980s, however, the market for consumer display holograms failed to develop, owing largely to further imaging constraints. Embossed reflective holograms are critically dependent on surface flatness; when mounted on magazine pages or even under glass, image and colour distortions are notoriously difficult to avoid. Second, the mirror-like surface has a disturbing tendency to reflect the light source to the viewers' eyes, being positively dangerous in sunlight. And third, a 'portable' hologram (that is, not permanently mounted in relation to an appropriate light source) inevitably becomes dim, fuzzy or distorted in typical room lighting.

When the original ambitions of inexpensive embossed holograms replacing silver-halide or DCG types proved impractical and unpalatable, developers quickly adapted. The 2D–3D hologram, merely an embossed hologram depicting a very shallow two- or three-plane image, was clearly inferior to the three-dimensional imaging possible with earlier forms, but proved ephemerally attractive for logos, children's stickers and

Table 11.1 Evolving cultural roles for holograms.

Widest availability	Application	Early developer or beneficiary	Cultural niche
1948–c1955	Holograms as intermediate elements in electron microscopy	Dennis Gabor	Microscopy
c1959–61	Holograms as 'wave photographs' and colour-sensitive optical elements	Yuri Denisyuk	Optical engineering
1959–c1990	Holograms as image processing devices	Willow Run Laboratories	Image processing
1965–	Holograms for motion / strain detection	Willow Run Laboratories	Mechanical engineering
(1966), 1981–c1995	Holograms as publication illustrations	Conductron, Light Impressions	Publishing
1967–	Holograms as memory devices	Willow Run Labs, Bell Telephone Labs	Data storage (experimental)
1968–72	Small holograms as promotional items	Conductron, McDonnell-Douglas, Light Impressions	Advertising
1968–c1990	(Pulsed) holograms as advertising media	Conductron, embossing firms	Advertising
1968–	Holograms as art media	Artists	Fine art
1970s–	Holograms as optical elements (HOEs)	Military aircraft designers	Aerospace
1973–c1980	Animated holograms	Multiplex Company	Consumer culture; films
1970s–	(Fictional) holograms as expressions of the future	Science fiction novels, cinema, television, videogames; conspiracy theorists	Popular culture; futurism
c1977–90	Holograms as home decoration	Numerous entrepreneurs	Consumer culture
1980–c2000	(Pulsed) holographic portraits	Artists, entrepreneurs	Contemporary history
1981–	(Embossed) holograms for packaging	Spectratek	Consumer goods
1981–c1990s	(Embossed) holograms for magazine advertising and promotional items	Various firms	Advertising
1983–	(Embossed) holograms for security	American Banknote Holographics, De La Rue Industries	Authenticating high-value products

(continued)

Table 11.1 Continued

Widest availability	Application	Early developer or beneficiary	Cultural niche
1990s	(Faux) hologram entertainers	Disney; Musion; AV Concepts	Theme parks, entertainment events
2000s	(Faux) hologram remote communications	CNN, video-link and telepresence firms	Current affairs, business communication
2010s	(Faux) hologram device displays	Nintendo 3DS, Microsoft HoloLens	Augmented reality for user interfaces

a limited class of magazine advertising. American Banknote's application of credit card security patches followed a similar route. The first MasterCard holograms were embossed holograms displaying a tiny statue of an American eagle, but the company later moved toward more complex multi-channel holograms that revealed different image elements depending on viewing angle. By the late 1980s, holograms could be generated by computer calculation, and so synthesized images began to dominate security applications. The computer-generated holograms could display more detailed images, but increasingly confined them close to the image plane to ensure ready visibility in all lighting.

These trends in recording technology, viewer acceptance and commercial goals led to a collapse of the holographic image back to the image plane and, indeed, to a replacement of imagery itself by eye-catching patterns. Embossed holograms sabotaged the more expensive and aesthetically oriented holograms, only to evolve into a security medium scarcely identifiable as holograms according to the original inventors' understandings. By the 2010s, the technology of hologram creation had been embodied and standardized in software and processing machines. Gone was the need for holographers or for discerning viewers.

Embossed holograms, with or without digital origination of the images, proved to be a zero-sum game. The production of holograms grew by at least a thousandfold between 1980 and 2010, but consumers gazed at them proportionately less frequently. Image-bearing holograms became rare, swamped by the increasingly ubiquitous diffractive reflectors.

All this, however, was behind the scenes for consumers. The rising sophistication of digital holograms on credit cards and tax stamps was little noticed. In fact, some counterfeiters, realizing that purchasers no longer knew what to expect, generated low-tech facsimiles of holograms that used mere foil and ink to simulate their bright reflections. Security firms could argue that their holograms now incorporated 'forensic' features—subtle, often microscopic details that could be revealed by those in the know—but secure authentication had left behind many businesses, shopkeepers and the public. Ironically, the wondrously realistic imaging medium revealed in 1964 had, fifty years later, once again become a technology characterized by secrecy.

Yet embossed holograms triggered an avalanche: they moved the state of the art from the spotlight of high culture (fine art and technical optimism) to low culture (children's products and kitsch decoration) and then to *no* culture (subliminally-acknowledged labels and invisible information). This transition almost entirely eradicated the short-lived art medium and its audiences. The stunning laser-lit holograms of the 1960s had been experienced by the lucky few of a single generation; the animated Multiplex holograms of San Francisco, a mass experience in popular culture during the mid-1970s, had disappeared by the end of the century. The speed of the transition discouraged attempts to preserve and archive the technology that had channelled magic and generated shock; the technical communities, responsive to wider cultural judgements, also diminished. As objects of curiosity, art and even decoration, holograms had a much shorter run than stereoscopes, and scarcely longer than 3-D movies. Much like the graphic arts between the wars, the popularization of holograms characterized a generation.

11.3 POPULAR AND UNPOPULAR CULTURES

This book has been the story of a visual aesthetic shaped by a series of new technologies and by professional, enthusiast and consumer cultures through the twentieth century. I have argued that holograms, in their multiple forms, became the channels for these cultural currents.

Like all cultural studies, these shifting targets and disparate perspectives have challenged our categories, confidences and certainties. What links a Leith–Upatnieks laser-lit hologram to a credit-card security sticker, or with the *Star Trek* holodeck? What unites audiences for consumer products, fine art and science fiction? And, most fundamentally of all, why do we expect these technologies to keep getting better?

The simple answer, illustrated through the book, is that audiences for visual media had distinct visions. Distinct subcultures were drawn to holograms for dissimilar reasons. Their creators—principally engineers, entrepreneurs, enthusiasts and artists—were motivated by a variety of aims and creative impulses. Consumers were attracted by wider themes grounded in the appeals of modernism and novel visual experience.

The optical effects produced by modern surface-relief holograms for security purposes bear little resemblance even to the embossed holograms of the early 1980s. Today those applications are less commonly seen than they once were; indeed, optically-produced holograms are sometimes described as 'classical' holography. Of course, such a label is in the eye of the creator: most holographers of the 1960s probably would prefer their deep, laser-lit holograms to the multicolour rainbow holograms of the 1970s. Their successors might prefer the embossed magazine covers of the 1980s to the shallow identity card holograms of the early 2000s. Different cultural groups with dissimilar backgrounds saw distinct possibilities for the medium of holography.

In cultural terms, this has nevertheless been a medium of middling impact. The numbers of hologram creators can be roughly estimated. From its origin in 1947 to 2015, some 20,000 authors have written at least one paper, report or book on holography. Nearly all of that activity occurred after 1965 as news of three-dimensional holograms spread. In the subsequent half-century, nearly all publications were by engineers and scientists, and around a thousand PhDs have been based on hologram research. Founded on some ten thousand patents, at least several thousand companies have entered the hologram market. At the narrow end of the tail, amateur hologram enthusiasts—estimated through participation in clubs and internet newsgroups—have been fairly static since the late 1990s and number around one thousand world-wide. And a similar number of exhibitions, big and small, have been mounted, involving several hundred or more active artists.[32]

But the consumers of holograms are more difficult to trace. We can grasp the scale and ephemeral nature of the audience through exhibition attendances. Tens of thousands attended some of the larger exhibitions, and a dozen or so long-term installations attracted an estimated hundreds of thousands per year. Integrated over the heyday of hologram shows, this suggests a total audience in the tens of millions. On the other hand, the sales of display holograms can be also estimated in orders of magnitude, and amount to between one and ten million at best.[33] Tellingly, relatively few have been displayed by their purchasers: eBay lists far fewer display holograms than vintage radios, for example.[34] Thus the exposure of viewers to holograms follows a familiar pattern: large audiences briefly or periodically were entertained by them, with progressively smaller numbers engaged over longer periods.

These are relatively small numbers in popular culture. For comparison, the number of regular television viewers in the USA is some 300 million, and the peak radio audience during the 1940s was about 100 million; the number of twentieth-century camera enthusiasts—buying magazines, film and accessories regularly—was of the order of ten million; the number of amateur telescope makers was as high as fifty thousand before and during the Second World War, and ham radio enthusiasts peaked at around a hundred thousand after it.

This crude accounting cannot assess the cultural impact of holograms, but it does reveal the 'elephant in the room': that non-viewers and non-enthusiasts of holograms vastly outnumber those who have engaged directly with them. It is ironic that most cultural exposure to holograms has been second- or third-hand: global audiences have

[32] Johnston, S. F., *Holographic Visions: A History of New Science* (Oxford: Oxford University Press, 2006), Appendix, and subsequent compilations over the decade since publication.

[33] Inferred from sales data provided by Gary Zellerbach for his seminal holography retail company, Holos Gallery (SFJ collection).

[34] Ross, J. to SFJ, telephone interview, 1 Aug 2014, London, SFJ collection.

been vicarious, experiencing holograms mainly via news stories, online dialogues and videogames. Holograms, unlike television or stereoscopes, did not become a technology of the home, but the *idea* of the hologram proved perennially seductive. The technology and experience of holograms have become entwined in contemporary legends and fantasy.

There is a related dimension touched upon in this account: the public/hidden axis of culture. Most activity in the hologram industries has been relatively invisible. As discussed in Section 4.4, the development of holograms during the late 1950s and early 1960s at the Willow Run Laboratories in Michigan and the Vavilov Optical Institute in Leningrad were for military applications.[35] The first generation of applications at Conductron, IBM and other corporations were sponsored by military contracts. Significant commercial activity during the 1970s moved to the aviation industry, where head-up displays and holographic testing of components was profitable but little-seen. And, as discussed in Chapter 10, commercial profits during the 1980s were again focused on covert activities: the creation of proprietary anti-counterfeiting measures and the prevention of fraud through intricately encoded security holograms. These secretive practices have bracketed the history of holograms, leaving a brief window during the 1970s and 1980s during which wider culture engaged directly with the technology. Their hidden history has left a legacy of popular intrigue surrounding holograms—a not-uncommon aftermath of technologies developed during the Cold War.[36]

The cultural history of holograms consequently is iceberg-like: much of it invisible to popular culture (and consequently much mythologized in fiction and conspiracy literature) but punctuated by episodes of visual wonder and public enchantment.

11.4 MAGIC, METAPHORS AND MATERIALISM

This book has identified forces and contexts that shaped—and sometimes failed to influence—cultural engagement with holograms.

The early chapters described the evolving visual grammar of the preceding century and how it was influenced by particular inventions (photography, stereoscopes, printing technologies and cinema), how a changing toolbox of graphic arts evoked new forms of visual surprise and how these new media channelled the themes and ambitions of modernity to ever-wider audiences.

[35] Johnston, S. F., 'Der parallaktische Blick: Der militärische Ursprung der Holographie [The parallax view: the military foundations of holography]', in: S. Rieger and J. Schröter (eds.), *Das holografische Wissen* (Dortmund, Germany: Diaphane, 2009), pp 33–57.

[36] See, for example, Johnston, S. F., *The Neutron's Children: Nuclear Engineers and the Shaping of Identity* (Oxford: Oxford University Press, 2012), Section 8.2.

The account also tracked the socioeconomic dimensions of influence: the growing power of individuals to pursue technical enthusiasms and the role of publishers, commercial suppliers and eventually governments in promoting innovative technical expertise. The result, from mid-century, was a generation of enthusiasts primed to explore the expressive, artistic and technically progressive potential of holograms. For each subculture, the promise of holograms was a powerful driving force.[37]

The preceding chapters have explored a broad and varied cultural terrain. I have sought to demonstrate that holograms have had cultural threads that can be tracked individually. These threads—alternately densely and sparsely distributed, influential and ephemeral—have left a complex tracery on popular culture.

I have argued that, against contemporary expectations, holograms did not produce an inevitable step-change in culture, but instead were grounded in cultural understandings that had developed over the preceding century. Holograms contributed to a mutating visual vocabulary that highlighted novelty and spectacle, but these had ancient roots. They were assessed in terms of contemporary confidence in innovation, progress and consumption. And their imagined futures reflected wider cultural anticipations more than they did the potential of the technology itself. This is, in short, a story of cultural continuity rather than discontinuity.

During their first decade of popularity—roughly the mid-1960s to mid-1970s—holograms were predominantly an American export, mirroring experiences with other technologies and fashions during the postwar period (Section 6.3). For the first audiences, most often engineers and media representatives, the visual realism of Leith-Upatnieks holograms was their dominant attribute. These mimetic qualities suggested a faithfulness of reproduction that was startlingly unfamiliar, making holograms the latest technology to expand engineering powers and consumer satisfaction. They were, nevertheless, successors to a string of inventions that had provided similar shocks over the previous generation: television and stereophonic sound. For an earlier generation, the advent of sound recording and radio had revealed similar engineering vistas.

[37] See, for example, 'New camera operating without lens shows scientific promise', *New York Times International Edition*, 11 Dec 1963, 14; Gilman, L. C., 'The promise of holography', *Bell Telephones Magazine* 46 (1967): 15–8; Rajchman, J. A., 'Promise of optical memories', *Journal of Applied Physics* 41 (1970): 1376–83; LaMacchia, J. T., 'Holographic optical memories: promise and problems', in: B. J. Thompson and J. B. Develis (eds.), *Developments in Holography II* (Bellingham, WA: SPIE, 1971), pp. 51–3; Zyabrev, V. A., 'Holography: a promise for information science', *Nauchno Tekhnicheskaya Informatsiya, Seriya* (1971): 30–4; Brenden, B. B., 'Acoustical holography: some techniques of promise', *Materials Evaluation* 31 (1973): 40A; Schuyten, P. J., 'Technology: unmet promise of holography', *New York Times*, 24 Apr 1980: D1; Leith, E. N., 'Holography – the promise fulfilled', *Proceedings of SPIE* 532 (1985): 2–5; Uhrin, R., 'Crystals could unlock an old dream for data storage volume holographic storage employing photofractives promises a new era of images', *Photonics Spectra* 27 (1993): 87; 'Holographic storage promises high data density', *Laser Focus World* 32 (1996): 81–2, 84–5, 87–8, 90, 93; 'Breakthrough promises cheaper, faster chips – optical switches turn to holography', *Computer* 34 (2001): 3, 'Outlook – holography and "spintronics" promise more data density I/O chip architecture enables increased bandwidth', *Electronic Products* 44 (2001): 6.

As traced in the opening chapters, innovation in, and cultural engagement with, holograms closely paralleled earlier visual technologies. For audiences and creators, holograms satisfied desires that had been identified, trialled and nurtured over five generations. A few examples underline the conjunction of popular experiences.

The visual spectacle of the first Leith–Upatnieks holograms had resonances with Victorian optical toys (Section 2.3), interwar graphic arts (Section 3.2) and evolving cinema (Section 3.5). Perceptual surprise was also a shared feature with two other visual media that emerged during the 1960s: Op Art and psychedelic light shows (Section 8.1).

Overlapping and 'multi-channel' three-dimensional images (in which one scene is replaced by another by shifting the hologram viewer's position), were features explored by the first holographic artists and cottage industrialists, but rehearsed the disorientation produced by the Russian interwar artists with their perspective-play and photomontage (Section 3.2).

These prior experiences extended visual grammar, and confined holograms to a narrow field of novelty. The gradual evolution of colour imaging over the century had been labelled with descriptions of 'natural', 'genuine' and 'realistic' colour, leaving the developers of techniques for creating the latest generation of faithful holograms with the unsatisfying superlative 'ultra-realistic'.[38] Similarly, Stephen Benton's MIT team in the 1990s had dubbed their own technical culmination the 'ultragram'—a wide-angle synthesized hologram reminiscent of Cinerama and other immersive wide-screen processes (Section 3.5).[39] Abstract pseudocolour holograms, explored from the 1970s, evoked pleasures similar to those introduced by Disney's *Fantasia*. And holograms as an educational and advertising medium echoed the public claims and applications for stereoscopes and cinema (Sections 2.3 and 3.5), printed illustrations (Section 3.1) and television (Section 3.5).

In public exhibitions, the hologram was most successful as a vehicle for mystery, bafflement and incongruity. Spectators between 1964 and the early 1980s focused on the bewildering aspects: its reliance on a mysteriously speckly laser beam; the immaterial image hanging in space; the hologram's ability to generate the uncanny appearance of solidity from a featureless flat plate; and, its revelation of mysterious scenes in impossible spaces. These compelling magicians' tricks updated the experience for new, more discerning, viewers. Magic evokes childlike wonder, encouraging viewers to suspend critical evaluation; it is not surprising the new technologies tap into our collective faith in progress. So ingrained is this innocent attraction that the appeal of this visual magic swept across the cultural spectrum from popular audiences to postmodern theorists (Figure 11.3).

[38] Bjelkagen, H. I. and D. Brotherton-Ratcliffe, *Ultra-Realistic Imaging: Advanced Techniques in Analogue and Digital Colour Holography* (London: CRC Press, 2013).

[39] Halle, M. W., S. A. Benton, M. A. Klug and J. S. Underkoffler, 'Ultragram: a generalized holographic stereogram', *Practical Holography V* 1461 (1991): 132–41; Klug, M. A., M. W. Halle and P. M. Hubel, 'Full color Ultragrams', *Proceedings of SPIE* 1667 (1992): 110–9.

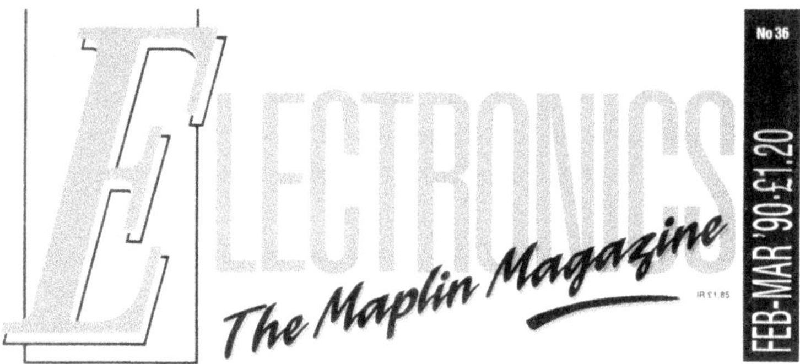

Fig. 11.3 Holograms to inspire multimedia hobbyists (*Electronics: The Maplin Magazine*, Feb / Mar 1990, © Maplin).

This too was familiar. Such experiences had for centuries been the domain of magic and theatre. The first laser-lit holograms rehearsed those baffling experiences and were susceptible to familiar explanations ('it's all done with mirrors'). Optical tricks in stage shows and cinema had deflated the impact of visual spectacle. Audiences' increasingly sophisticated visual education over the previous decades explains the muted impact of holograms in culture.

Visual surprise and spectacle dominated their public experience, but the growing accessibility of holograms further diluted their impact. Democratizing the visual arts traded off their sublimity. The environment for enjoying holograms during the 1960s was conditioned by the technologies of the previous century, which had brought mesmerizing individualistic experiences to mass audiences. In their own way, a chain of more personal optical innovations successively conditioned audiences and trained experts to refine their visual tastes to enjoy, and expect, new forms of visual dissonance. Along this trajectory—following the 'perspective play' of interwar graphic art, coincident with 1960s op art, but falling behind cultural expectations in the era of video games—the appeal of holograms can be recognized as a transitory phenomenon. Like those earlier optical experiences, holograms were rapidly absorbed, assessed and ultimately rejected by successive audiences.

Later variants of holograms mixed some of those earlier technologies back in to revive the thrill. The few seconds of motion created by integral holograms made the experience of viewing them much like the earlier zoetrope, but augmented by the novel sense of image depth. Holographic portraits were startlingly—and even uncomfortably—lifelike, producing responses from 1980s audiences probably similar to those of the 1850s seeing their first framed family photograph. The intricacy, shimmering colour and changing imagery of the hologram identity cards entranced their first viewers as Victorian optical toys had done, but held the promise of more sophisticated properties to discover. Each of these experiences separated by time and space was, at least for a while, exciting, novel and precisely aimed to extend the visual grammar of the age.

Yet the fleeting impact of holographic advertising and packaging illustrates this fickle appeal. Awe-inspiring scenes, unexpected motion and the surprise of colour vied successively for attention, and the imaging of visual depth repeatedly attracted fresh waves of viewers. Spectators acclimatized rapidly to the novelty of these experiences. From this perspective, the popular history of holograms looks familiar.

It is intriguing, then, that the greatest impact of holograms has been in our minds. Holograms have had their most enduring influence—from their public revelation in 1964 and over the following fifty years—as a metaphor for the future. This longevity is unusual, if not unique. Ironically, it may be because holograms have not yet convincingly been assimilated in a stable form within consumer culture. Represented as a technology that will eventually and inevitably appear, holograms continue to channel cultural dreams. By contrast, other technologies that had played this role have lost potency when they were achieved in practice. Television of the 2000s attained the large, flat-screen and even three-dimensional form predicted from the 1880s, rendering it banal; the gleaming rockets of 1930s science fiction were transformed into complex but vulnerable behemoths by the NASA and Soviet space programmes. More feasible than time-travel and more democratic than rocket-ships, holograms have channelled optimism of a better and more exciting future.

I have argued that these cultural themes co-mingled and interacted. Some—like the power of holograms to visually shock—resonated briefly. Others—particularly holograms as channels to the mystical and magical—tap into ancient and enduring cultural attractions. And still others—notably holograms as representative of the future—continue to grow in popularity and to accrete new imagined attributes.

Technology fertilizes dreams and aspirations, some of which are endemic in human cultures. Holograms did not merely create aspirations anew; their special power has lain in channelling the expression and realization of long-familiar themes.

APPENDIX: A TAXONOMY OF HOLOGRAMS

I have to start my classes with what holography isn't . . . you have to unwind it several layers . . . so many of my students assume that, if they've seen it in a movie, if they've seen it on television, then that's reality.[1]

Fred Unterseher, Santa Clara, CA, 2003

In terms of usage of the term 'hologram', I have tried to let it go at this point. You can really let it get to you . . . but we kind of lost that battle.[2]

Gary Zellerbach, 2014

The cultural identity of holograms is closely linked to their mystery, novelty and utility, i.e. to how they *behave* in particular environments. Their technical properties provide cues for the metaphors that describe them. At the risk of deflating their appeal, this Appendix distinguishes the principal varieties, gives a brief summary of their optical characteristics, and distinguishes them from other technologies.

A.1 REAL-WORLD HOLOGRAMS

Table A1 categorizes the characteristics of the principal varieties of hologram. As argued throughout this book, the gradual evolution of these variants, and the distinct contexts in which they developed, means that texts that describe them are products of their times. Practical books published from the 1980s, as visual properties began to stabilize, are among the better guides explaining the principles, recording and lighting of holograms.[3]

A.2 MISIDENTIFYING HOLOGRAMS

Because holograms themselves are so varied in their visual effects, other optical technologies can masquerade or be misidentified as holograms. Despite the plethora of claims by companies, promoters and bloggers, most of these pseudo-holographic techniques combine elements available long before holograms were invented, albeit combined in innovative ways in new contexts and for new audiences.

The display of three-dimensionality is one of the most widely recognized characteristics of holograms, although not all holograms do so. To confuse matters further, other technologies can produce some of the optical effects characteristic of three-dimensional images.

[1] Unterseher, F. to S. F. Johnston, interview, 22 Jan 2003, Santa Clara, CA, SFJ collection.

[2] Zellerbach, G. A. to S. F. Johnston, telephone interview, 12 Jun 2014, SFJ collection.

[3] See, for example, Unterseher, F., J. Hansen and B. Schlesinger, *Holography Handbook: Making Holograms the Easy Way* (Berkeley, CA: Ross Books, 1981); Saxby, G., *Practical Holography* (New York: Prentice-Hall International, 1987); Benton, S. A. and V. M. Bove Jr., *Holographic Imaging* (New Jersey: John Wiley & Sons, 2008).

Table A1 Varieties of hologram

	Hologram type	Lighting for recording/viewing	Image position	Image depth	Image properties	Typical subjects
1947–60	Gabor	Filtered monochromatic lamp	Focused centimetres from hologram	Fraction of millimetre	Requires viewing optics (e.g. microscope)	Line drawings and electron-microscope shadowgrams
1961	Denisyuk reflection	Filtered monochromatic lamp/white light	Against hologram plate	Fraction of millimetre	Usually tinted	Reflective objects, curved mirror
1961–	Leith–Upatnieks 2D transmission	Filtered monochromatic lamp	Focused centimetres from hologram	Fraction of millimetre	Requires viewing optics (e.g. microscope)	Line drawings, then photographic transparencies
1964–67	Leith–Upatnieks 3D transmission	Laser	Behind hologram plane	Up to several metres, but usually tens of centimetres	Full depth cues (binocular disparity, parallax, accommodation, convergence); sometimes granular appearance from laser speckle	Laboratory objects, chess pieces, knick-knacks
1965–	Holographic interferogram	Laser	Behind hologram plane	Up to several metres, but usually tens of centimetres	Full depth cues; sometimes granular appearance from laser speckle	Engineering scenes undergoing subtle movement or stress
1965–	Denisyuk reflection	Laser/white light	Behind hologram plate	Usually tens of centimetres	Full depth cues; tinted	Museum objects, still-lifes, aesthetic subjects
1966–72	Image-plane	Laser/white light	Straddling hologram plane	A few centimetres	Full depth cues, but fuzzy and colour-fringed further away from plate	Knick-knacks

1967–	Pulsed laser	Laser	Behind hologram plane	Up to a metre or so, but usually tens of centimetres	Full depth cues; sometimes granular appearance from laser speckle	Rapidly-moving engineering scenes, portraits, figure studies, erotica
c1970–	Multi-channel	Laser / white light	Behind or straddling hologram plane	Usually centimetres to tens of centimetres	Two or more distinct three-dimensional scenes, each visible at particular viewing angles *or* distinct illumination directions	Product advertising, models
1972–c1985	Benton rainbow transmission (or 'White Light Transmission, WLT')	Laser / white light	Behind or straddling hologram plane	Usually centimetres to tens of centimetres	Spectrally tinted along vertical axis; no vertical parallax; fuzzier away from hologram plane	Knick-knacks, aesthetic subjects
1973–c1978	Cross 'Multiplex' (integral hologram or holographic stereogram)	White light (for filmed scene); Laser (for hologram recording) / white light	Within a cylindrical hologram	Usually a few centimetres	Spectrally tinted along vertical axis; no vertical parallax; limited horizontal resolution	Filmed outdoor scenes, portraits, erotica
c1974–	Pseudocolour	Laser / white light	Behind or straddling hologram plane	Usually a few centimetres	Three or more overlapping colours	Abstracts, simple compositions
c1978–	Dichromated gelatine (DCG or 'dichromate') and photopolymer media	Laser / white light	Usually image-plane	Usually a few centimetres	Relatively bright; sometimes low contrast owing to light scatter; fuzzier away from hologram plane	Often pendant- to head-sized models or objects
c1980–	Embossed	Laser / white light	Straddling hologram plane	Usually a few centimetres	Very bright; often distorted by lack of flatness; spectrally tinted	Still-life subjects, security patterns
1981–c1985	McGrew 2D–3D (embossed)	Laser / white light	Straddling hologram plane	A few millimetres	Very bright; limited three-dimensionality	Advertising and children's stickers

(*continued*)

Table A1 Continued

	Hologram type	**Lighting for recording/viewing**	**Image position**	**Image depth**	**Image properties**	**Typical subjects**
c1967 but esp. post 1985	Computer-generated hologram (CGH)	Laser/white light	Usually straddling hologram plane	Usually centimetres to tens of centimetres	Very bright	Security and decorative patterns; display posters
1990s–	Holographic video	Various	Via computer interface	Usually centimetres to tens of centimetres	Varies	Electronic transmission and reconstruction of holographic image; limited demonstrations
1990s–	Haptic holograms	Various	In front of hologram plane; tactile sensation requires associated equipment	Usually centimetres to tens of centimetres	Varies	Computer-generated touch sensation; limited demonstrations

Visual depth perception relies on four distinct detection mechanisms of the eye and brain:

(a) *Binocular disparity*: the difference between the images seen by each eye owing to its unique position. Objects near the observer will have distinctly different angles with respect to the two eyes, while distant objects will not. The result is that the image seen by each eye will include objects having different relative positions depending on their respective distances.
(b) *Parallax*: the sensing of this disparity in the apparent position of viewed objects when the scene is viewed from a different position. It is most obvious when one viewed object blocks another. Vertical parallax is observable only by moving the viewpoint vertically, e.g. by bobbing the head up and down.
(c) *Accommodation*: the adjusting of the muscles of the eye to focus on a distant object. When viewing a photograph or television screen—even a stereoscopic photograph or 3D television image—all the objects in the recorded scene are focused by the eye at the same distance: they are either sharp or blurry depending on the setting of the lens that recorded the image, and how well the eye focuses on the photographic paper or screen pixels.
(d) *Convergence*: the rotation of the two eyes in opposite directions to centre the object of interest on the retinas.

A.2.1 Stereoscopic imagery

As discussed in Section 2.3, the oldest form of three-dimensional imaging relies on supplying just two versions of a three-dimensional scene, one to each eye. This will give the perception of three-dimensionality, but only from a single fixed perspective. The most common forms of stereoscopic imaging do not provide parallax (which allows the viewer to 'look behind' a foreground object) or visual accommodation (requiring the viewer to focus at different depths of the scene).

The process of visual depth perception by binocular viewing, dubbed stereopsis, was first described by Charles Wheatstone (1802–75) in 1838, and the first practical stereoscope was invented by David Brewster (1781–1868) in 1849. In Brewster's stereoscope, the perception of depth relies only on binocular disparity, and the viewer's eyes fuse the two images by rotating (the action of convergence). The other stereoscopic cues of parallax and accommodation are not involved. As a result, the actions of focusing and head movement are decoupled from eye convergence. When the disparity between these cues is too extreme, as in poorly designed 3D movies, the viewer will experience eye fatigue or disorientation.

In their most limited form, stereoscopic displays provide a separate image to each eye for a single observer. The Victorian stereoscope, based on Brewster's design, allowed the observer to view a card on which two photographs of a scene were positioned side-by-side. The photographs were recorded by camera lenses separated by the same distance as the eyes themselves.[4]

Later examples of stereoscopic technology are optical schemes that restrict the images visible to each eye in other ways. For motion pictures and television, two techniques became popular. The first employs *anaglyphs*: overlapping images projected in two colours, one for each eye. By wearing viewing glasses having coloured filters that block the opposite colour, each eye sees only one of the two images, and so the viewer perceives a three-dimensional image from the separate views. The same technique can be used for printed material, too, such as comic book

[4] Holmes, O. W., 'The stereoscope and the stereograph', *The Atlantic Monthly* III (1859): 738–48. If the distance between lenses is increased, the perception of depth is magnified. This technique became central to stereoscopic photogrammetry, in which photographs from airborne cameras are taken several hundred feet apart as the aircraft moves, to measure the height of terrain and features.

illustrations employing two coloured inks. The second popular technique is to employ polarizing filters (typically the Polaroid material patented by Edwin Land in 1929): the two projected images pass through separate polarizers with their polarizing axes opposed, and are reflected from a screen that preserves this polarization. When viewed through the observer's glasses, which contain similarly oriented polarizers, a different image is seen by each eye, and stereopsis again occurs.[5]

An extension of this general class of stereoscopic imaging is virtual reality (VR) technologies that employ a viewing helmet with a separate digital display for each eye and a motion-detection system to detect the orientation of the observer and to alter the calculated images accordingly. VR systems have the advantage of being able to recalculate the stereoscopic images on the fly, allowing for both parallax and binocular disparity. The first consumer devices of the 1990s proved unpopular, in part, because of unsophisticated graphics and time lags in the display calculations. The delay between the viewers' movements and displayed image could cause disorientation and queasiness. The goggles also obscured the local environment, isolating the viewer. Rather than enhancing reality they substituted a noticeably inferior virtual alternative.

Stereoscopic images of this kind are seldom mistaken for holograms because the technology is so clearly linked to the individual viewer. However, as carefully engineered 3D cinema and television programmes have demonstrated, audiences can find the technology compelling and immersive.

A.2.2 Auto-stereoscopic imagery

An alternative to head-mounted apparatus is an auto-stereoscopic display. Such a display allows the observer to perceive depth without glasses. However, a carefully designed optical arrangement is still required to provide distinct views to each eye. The two most common techniques are lenticular surfaces and parallax barriers.

A lenticular surface is one that either incorporates small lenses or other optical elements to restrict the view to a narrow range of angles. In this way, each eye can be limited to a different view. The first American patent for lenticular photography envisaged cylindrical lenses arranged side by side.[6] One angle of view through the cylindrical lens channels the left-eye image, and the other channels the right-eye image. Lenticular images can be manufactured as two-dimensional images consisting of interleaved strips of the left- and right-eye images, combined with a moulded plastic overlay that incorporates a fine grating of cylindrical lenses or prisms. Alternatively, such printed images can be mechanically corrugated to achieve the same result. The linear ridges make lenticulars readily identifiable by touch.

Lenticular products were marketed by the New York-based Vari-Vue Company from the early 1940s in the USA, and during the 1960s they were licensed internationally. The products included political campaign buttons, postcards, children's product prizes, wall hangings and even billboards. Most of their early products featured two distinct images, such as a winking eye or other 'flicker animation', rather than three-dimensional views, however, a technique revived in some 2D–3D holograms.

The heyday for such lenticular products was the 1960s, so there was some overlap and cross-fertilization with early holography. *Look* magazine, for example, included a black and white lenticular picture of Thomas Edison and, a few months later, a colour lenticular advertisement for a

[5] Norling, J. A., 'The stereoscopic art', *Journal of the Society of Motion Picture and Television Engineers* 60 (1953): 268–308.

[6] Coffey, D. F. W., Pat. No. US 2,063,985 A 'Apparatus for making a composite stereograph' (1936), assigned to Winnik Stereoscopic Processes Inc.

Division of the Eastman Chemical Company in 1964, the year that three-dimensional laser holograms were announced.[7]

Variants of this scheme (used, for example, in the Nintendo 3DS hand-held game console, 2010) employ other forms of parallax barrier. These are typically opaque masks manufactured and positioned over the strip image to restrict the viewpoint of each eye, rather like looking through a picket fence at an object close behind it. As with embossed fine prisms and mechanical corrugation, parallax barriers require precise manufacture and are designed for a specific viewing position.

Such lenticular displays are well-suited to digital imaging for hand-held and larger display devices because the required stripe displays can be produced by software, and supplemented by a plastic overlay consisting of the cylindrical grating. In such systems, the observer's viewing position is usually limited. If the observer is too far off-axis, the viewing geometry will no longer allow each eye to see its intended image and the perception of depth will be distorted or fail. None of the common varieties of lenticular technology allows for vertical parallax: if the viewer's head is tilted 90 degrees, the perception of three-dimensionality disappears. As a result of these optical and geometrical requirements, lenticular imaging has a perceptibly lower resolution than a regular photograph, cinema image or hologram.

An earlier and more sophisticated form of lenticular imaging employs microscopic lenses rather than prisms or corrugations. The *fly's eye lens* is an array of small lenses, often embossed on plastic, which can individually view a scene and record it as an array of miniature photographs. When this image is subsequently viewed or projected through the fly's eye lens the viewer will be able to change position to gain a sensation of the depth of the scene. The scheme, dubbed *integral photography*, was introduced by physicist Gabriel Lippmann in 1908.[8] The technique allows for both horizontal and vertical parallax as well as accommodation. In practice, the integral photograph acts like a window beyond which the scene is visible, and so resembles a Leith–Upatnieks or Denisyuk hologram. However, unless a very fine microlens array is used (which introduces its own optical and manufacturing problems) the resolution is noticeably poorer than a hologram. The small optical elements reduce the image resolution by diffraction, and make the image appear grainy or pixelated owing to the limited number of lenses that can be used.

Audiences nevertheless link lenticular technologies with holography; the popular understandings and markets overlap so closely that some holography firms, such as Spatial Imaging (London) have shifted the brunt of their marketing and production to lenticulars.

A.2.3 Volumetric imagery

Volumetric displays generate a three-dimensional image in a volume of space that can be viewed from different angles. In this sense, it is a genuinely three-dimensional technique that makes use of the four cues of binocular disparity, parallax, accommodation and convergence. The popular notion of such imagery is close to the depiction of the hologram of Princess Leia in *Star Wars*, although the film erroneously suggests that this is achieved from a single projector, and that the process is holographic.

This perception of depth can be accomplished either by projecting the image point onto a medium that reflects it, or by generating an image point from a light source. One example of the technique is to project two-dimensional images from multiple projectors onto a column of fine mist or spray. The droplets will reflect the image from two or more projectors back to the eyes of

[7] *Look* magazine 25 Feb 1964, p. 102; *Look* magazine 7 Apr 1964, p. 89.

[8] Lippmann, G., 'Épreuves réversibles. Photographies intégrales', *Comptes Rendus de l'Académie des Sciences* 146 (1908): 446–51.

the observer. The projected images may be either analogue (e.g. from movie film) or digital (e.g. from LCD projectors). The limitation—as with genuine holograms—is that the viewing volume is relatively small.

For the bright-point method, best suited to digital imagery, one technique is to employ a rotating panel consisting of an array of LEDs, with the LED brightness synchronized to the rotation to sweep out a volumetric image. With sufficiently rapid rotation, a three-dimensional scene will be perceived by the observer. A more elaborate alternative is to cause points in a transparent solid, liquid or gas to emit light, for example by focusing a scanned laser beam in a fluorescent medium. The bright-point method, either using a static-volume or swept-volume technique, is again limited to a relatively small cylindrical space, and image resolution will be determined by the density of LED light sources.

For viewers, the experience is similar to viewing a Cross Multiplex hologram of the 1970s, but with the added appeal of dynamically displaying computer- or video-generated imagery. On the other hand, viewers cannot interact with these mechanically-scanned or laser-scintillated images. Volumetric displays are relatively complex and limited in physical size. Just as with Multiplex holograms, the dedication of the necessary space has discouraged home use.

A.2.4 Video techniques

Contemporary audiences have also been misled by novel video techniques inaccurately identified as holographic. In 2008, CNN touted its introduction of 'holograms' in its broadcasts with a visual effect that recalled the Princess Leia scene in *Star Wars*. Its reporter in a remote studio spoke 'via hologram' to the anchor man as the camera panned around the set. Despite the sophisticated synchronization of camera movements on both sets, however, the effect was merely an extension of existing video technology. The reporter was imaged by encircling video cameras, and their positions, movements and optical settings were synchronized with cameras on the main set so that the appropriate remote camera image was selected. The two studio images were overlayed by conventional green-screen video techniques. It should be noted that—as with the Pepper's Ghost technique described below—television viewers saw the realistic composite image, but the live presenters were not able to see each other except on studio monitors.

More generally, *telepresence* techniques extend two-way videoconferencing by improving the sense of co-location (e.g. by employing better display technology) or the variety of tasks that can be performed (e.g. to allow collaborative software development or repairing a fault). Rather than being an illusion portrayed for immobile audiences, telepresence provides an enhanced visual experience for participants in two or more locations, but it is neither inherently three-dimensional nor holographic.

A.2.5 Pepper's Ghost

The visual trick most often misidentified with holograms is Pepper's Ghost, a spectacular stage effect first seen by Victorian theatre audiences. An off-stage actor, illuminated by a bright lamp, would be visible to the audience via his reflection in an unseen pane of glass on stage. The ghost illusion had been conceived around 1858 by Henry Dircks, a retired civil engineer, possibly with the intention of educating the public about the conjuring tricks of spiritualist charlatans. His scheme was developed and first employed by analytical chemist John Pepper for stage plays from 1862 and patented in 1879.[9]

[9] On the disputed origins of the invention, see Pepper, J. H., *The True History of the Ghost: And all about Metempsychosis* (London: Cassell, 1890). A Renaissance precursor for the illusion, however, is described in Della Porta, G. B., *Magiæ naturalis sive de miraculis rerum naturalium* (Napoli: 1558), a compendium of magicians' tricks.

Fig. A1 Pepper's Ghost stage illusion (Marion, F., *L'Optique* (1869), Fig. 73).

With a live actor, the reflected image was three-dimensional and startling, and could appear to hang in space if the reflector were suitably positioned (Figure A1). Note that, unlike the artist's rendition, the actor on stage was not able see the apparition: it was visible only to the audience via the reflection in the glass. For late-Victorian audiences, a magic lantern slide became a simpler and less costly alternative, although in this case the reflected image was, like the slide itself, two-dimensional. Variations of the trick were popular with a generation of theatre and fairground audiences before the introduction of motion pictures.

This trick has been popular in theme parks and stage spectacles since the 1960s, where the audience members cannot change position enough to detect the two-dimensionality of the projected images. The technique has been revived with images projected from video sources rather than live actors, and with alternatives to the original glass pane reflector. One substitution is to employ a thin reflective film, a water spray mist or fog curtain or even a white mannequin as a screen on which to project the video image. The mannequin is especially well suited to close viewing by audiences who can get a strong impression of three-dimensionality as the projected image 'wraps around' the figure. The key to the illusion is to ensure that the audience cannot detect the screen itself, which usually is accomplished by arranging a dark background and ensuring that the reflector or screen is unlit by other sources.

It is worth noting that most implementations of Pepper's Ghost have none of the features of three dimensional imaging: they require neither binocular disparity, motion parallax, accommodation nor convergence. Observers use these visual cues merely to locate the position of the image in space as they would when viewing a cinema or television screen. But the illusion is more marked when viewers can see other objects nearby—a live performer or furniture—at different distances from the (invisible) screen. Tricked into perceiving the spectre as a living person rather than a projected image, their brains 'fill in' the missing dimension of depth.

Examples include Disney attractions such as the 'Haunted Mansion' (opened in Anaheim, 1969, with large plate glass reflectors)[10] and 'Pirates of the Caribbean' (updated to include projection onto a curtain of mist in 2006), 'The Wizarding World of Harry Potter' (Universal, Orlando, opened 2010), the Hatsuni Miku 'virtual star' in Japan (2010) and the Tupac Shakur 'hologram' (Coachella festival, 2012), the latter three using a metallized Mylar screen as reflector. Following the Coachella event in particular, popular identification of virtual stage performances as 'holograms' dominated web traffic.[11]

In short: entertainment 'holograms' are novel embodiments of a long-standing stage technique. Conventional head-up displays and teleprompters employ the same principle.

[10] Sweezey, C. O., 'All that's 3D is not holography: Disney World's Haunted House', *Holosphere*, 10 (1), Jan 1981: 1.

[11] For a commercial version, see Maass, U., Pat. No. US5865519 (A) 1999–02-02 'Device for displaying moving images in the background of a stage' (1999), assigned to Musion. In more sophisticated implementations, the two-dimensional video images can be generated by compositing (combining image elements from) film or video footage and rotoscoping (tracing a recorded image for substitution of a body double or an inset of computer generated imagery (CGI)).

BIBLIOGRAPHY

ARCHIVES CONSULTED

Publicly accessible collections

ACNMAH	Archives Center of the National Museum of American History, Washington, DC
IC	Imperial College Archives, London
Sci Mus	Science Museum, London
SIA	Smithsonian Institution Archives, Washington, DC

Private collections

Charnetski collection	Clark Charnetski, Ann Arbor, MI, USA
Funkhouser collection	Arthur Funkhouser, Bern, Switzerland
Naeve collection	Ambjörn Naeve, Stockholm, Sweden
Richardson collection	Martin Richardson, Leicester, UK
Ross collection	Jonathan Ross, London, UK
SFJ collection	Sean F. Johnston, Dumfries, UK
Zellerbach collection	Gary Zellerbach, San Francisco, USA

BOOKS AND DISSERTATIONS

Kodak Data Book on Copying (Rochester, NY: Eastman Kodak Co., 1941)

Kodak Master Photoguide (Rochester, NY: Eastman Kodak Co., 1951)

Les Prix Nobel En 1971 (Stockholm: Imprimerie Royale P. A. Norstedt & Sonner, 1971)

Agar, J., *The Government Machine: A Revolutionary History of the Computer* (Cambridge, MA: MIT Press, 2003)

Arcangeli, A., *Cultural History: A Concise Introduction* (New York: Routledge, 2012)

Ashley, M., *The Gernsback Days* (Holicong, PA: Wildside Press, 2004)

Baez, A. V. and T. H. Jeong, *Introduction to Holography* (Encyclopaedia Britannica Educational Corp., 1972)

Batchen, G., *Burning with Desire: The Conception of Photography* (Cambridge, MA: MIT Press, 1999)

Baudrillard, J. and S. F. Glaser (transl.), *Simulacra and Simulation* (Ann Arbor: University of Michigan Press, 1981)

Behrens, R. R. (ed.), *Ship Shape, a Dazzle Camouflage Sourcebook: An Anthology of Writings about Ship Camouflage during World War I* (London: Bobolink, 2012)

Benjamin, W. and A. Underwood (transl.), *The Work of Art in the Age of Mechanical Reproduction* (New York: Penguin, 2008)

Benton, S. A. and V. M. Bove Jr., *Holographic Imaging* (Hoboken, NJ: John Wiley & Sons, 2008)

Bijker, W. E., T. P. Hughes and T. J. Pinch, *The Social Construction of Technological Systems: New Directions in the Sociology and History of Technology* (Cambridge, MA: MIT Press, 1990)

Bjelkhagen, H. I., *Holography and Philately: Postage Stamps with Holograms* (Dartford, UK: XLibrisUK, 2014)

Bjelkagen, H. I. and D. Brotherton-Ratcliffe, *Ultra-Realistic Imaging: Advanced Techniques in Analogue and Digital Colour Holography* (London: CRC Press, 2013)

Blackmore, S., *The Meme Machine* (Oxford: Oxford University Press, 1999)

Bloom, M. T., *The Brotherhood of Money: the Secret World of Bank Note Printers* (Port Clinton, OH: BNR Press, 1983)

Born, M. and E. Wolf, *Principles of Optics: Electromagnetic Theory of Propagation, Interference, and Diffraction of Light* (London: Pergamon Press, 1959)

Bothamley, C. H., *Ilford Manual of Photography* (London: Brittania Works Co., 1890)

Bowler, P. J., *Science for All: The Popularization of Science in Early Twentieth-Century Britain* (Chicago: University of Chicago Press, 2009)

Brand, S., *Whole Earth Catalog—Access to Tools* (Menlo Park, CA: Portola Institute, 1970)

Braunstein, P. and M. W. Doyle, *Imagine Nation: the American Counterculture of the 1960s and '70s* (London: Routledge, 2002)

Brewster, D., *The Stereoscope: Its History, Theory and Construction, with its Application to the Fine and Useful Arts and to Education* (Hastings on Hudson, NY: Morgan & Morgan, 1856)

Bromberg, J. L., *The Laser in America, 1950–1970* (Cambridge, MA: MIT Press, 1991)

Brunner, J., *Stand on Zanzibar* (London: Doubleday, 1968)

Bud, R., *The Uses of Life: A History of Biotechnology* (Cambridge: Cambridge University Press, 1993)

Burbank, W. H., *The Photographic Negative—written as a practical guide to the preparation of sensitive surfaces by the calotype, albumen, collodion, and gelatin processes, on glass and paper, with supplementary chapters on development, etc* (New York: Scoville Manufacturing Co, 1888)

Burns, R., *John Logie Baird, Television Pioneer* (Stevenage: Institution of Electrical Engineers, 2000)

Butterworth, H., *The Science and Art Department, 1853–1900*, PhD thesis, Sheffield (1968)

Cardwell, D. S. L., *The Organisation of Science in England* (London: Heinemann, 1972)

Carlson, S., *Photo-Kinks: The Photographer's Handy Book* (Minneapolis, MN: Huddle, 1937)

Carson, R., *Silent Spring* (Boston: Houghton-Mifflin, 1962)

Clarke, A. C., *The Fountains of Paradise* (New York: Ballantine, 1978)

Clayton, R. J. and J. Algar, *The GEC Research Laboratories 1919–1984* (London: Institution of Engineering and Technology, 1989)

Club of Rome, *The Limits to Growth: A Report for the Club of Rome's Project on the Predicament of Mankind* (New York: Universe, 1972)

Coe, B. and P. Gates, *The Snapshot Photograph: The Rise of Popular Photography, 1888–1939* (London: Ash & Grant, 1977)

Cohen, A. J., *Imagining the Unimaginable: World War, Modern Art, and the Politics of Public Culture in Russia, 1914–1917* (Lincoln, NE: University of Nebraska Press, 2008)

Connerton, P., *How Societies Remember* (Cambridge: Cambridge University Press, 1989)

Corn, J. (ed.), *Imagining Tomorrow: History, Technology and the American Future* (Cambridge, MA: MIT Press, 1986)

Corn, J. J., *User Unfriendly: Consumer Struggles with Personal Technologies from Clocks and Sewing Machines to Cars and Computers* (Baltimore: Johns Hopkins University Press, 2011)

Crary, J., *Techniques of the Observer: On Vision and Modernity in the Nineteenth Century* (Cambridge, MA: MIT Press, 1990)

Cullingham, G. G., *F. J. Camm, The Practical Man* (Windsor, UK: Thamesweb Publishing, 1996)

Daston, L. (ed.), *Biographies of Scientific Objects* (Chicago: University of Chicago Press, 2000)

Davies, O., *Witchcraft, Magic and Culture, 1736–1951* (Manchester: Manchester University Press, 1999)

Davis, W. S., *Practical Amateur Photography* (Ann Arbor, MI: Little, Brown & Co., 1923)

Dawkins, R., *The Selfish Gene* (Oxford: Oxford University Press, 1976)

De Solla Price, D., *Little Science, Big Science* (Washington, DC: Columbia University Press, 1963)

DeFreitas, F., A. Rhody and S. W. Michael, *Shoebox Holography: A Step-By-Step Guide to Making Holograms Using Inexpensive Semiconductor Diode Lasers* (Berkeley, CA: Ross Books, 2000)

DeLeskie, P., *The Underwood Stereograph Travel System: a Historical and Cultural Analysis*, MA thesis, Concordia University (2001)

Della Porta, G. B., *Magiæ naturalis sive de miraculis rerum naturalium* (Napoli: 1558)

Dick, P. K., *A Scanner Darkly* (New York: Doubleday, 1977)

Dizer, J., *Tom Swift & Company: 'Boys' Books' by Stratemeyer and Others* (Jefferson, NC: McFarland, 1982)

Dowbenko, G., *Homegrown Holography* (Garden City, NY: Amphoto, 1978)

Duncan, J., *A Companion to Cultural Ecology* (New York: Wiley-Blackwell, 2007)

Dupree, A. H., *Science in the Federal Government: A History of Policies and Activities* (Baltimore: Johns Hopkins University Press, 1986)

Eastman Kodak Co., *Kodakery: A Magazine for Amateur Photographers* (Rochester, NY: Eastman Kodak Co., 1913)

Eastman Kodak Co., *Photography: An Outline Course for Instructors* (Rochester, NY: Eastman Kodak Co., 1926)

Eco, U. and W. Weaver, *Travels in Hyper-Reality: Essays* (London: Picador, 1986)

Eichert, E. S. and A. H. Frey, *Holography in Driver Education, Training, Testing, and Research* (Springfield, VA: National Highway Traffic Safety Administration, 1978)

Elkins, J., *Visual Studies: A Skeptical Introduction* (New York: Routledge, 2003)

Ellison, W. F. A., *The Amateur's Telescope* (Belfast: Carswell & Son, 1920)

Ellul, J., *The Technological Society* (New York: Knopf, 1964)

Engel, J., *Gene Roddenberry: The Myth and the Man behind Star Trek* (New York: Hyperion, 1994)

Erlich, P. R., *The Population Bomb* (New York: Ballantyne, 1968)

Fara, P., *An Entertainment for Angels: Electricity in the Enlightenment* (Cambridge: Cambridge University Press, 2002)

Fisher, M., *Capitalist Realism: Is There No Alternative?* (Ropley, UK: Zero Books, 2009)

Ford, B. J., *Images of Science* (New York: Oxford University Press, 1993)

Frayling, C., *Mad, Bad or Dangerous? The Scientist and the Cinema* (London: Reaktion, 2005)

Friedman, J. S., *History of Color Photography* (London: Focal Press, 1972)

Fuller, B., *Operating Manual for Spaceship Earth* (New York: E. P. Dutton & Co., 1968)

Fuller, B. and E. J. Applewhite, *Synergetics: Explorations in the Geometry of Thinking* (New York: Prentice Hall, 1982)

Gabor, D., *The Electron Microscope: Its Development, Present Performance, and Future Possibilities* (London: Electronic Engineering, 1948)

Gabor, D., *Inventing the Future* (London: Secker & Warburg, 1963)

Galison, P., *Image and Logic: A Material Culture of Microphysics* (Chicago: University of Chicago Press, 1997)

Galison, P. and B. Hevley (eds.), *Big Science: The Growth of Large-Scale Research* (Stanford, CA: Stanford University Press, 1992)

Geertz, C., *The Interpretation of Cultures: Selected Essays* (New York: Basic Books, 1973)

Gelber, S. M., *Hobbies: Leisure and the Culture of Work in America* (New York: Columbia University Press, 1999)

Gernsheim, H. and A. Gernsheim, *The History of Photography* (Oxford: Oxford University Press, 1955)

Gerovitch, S., *From Newspeak to Cyberspeak: A History of Soviet Cybernetics* (Cambridge, MA: MIT Press, 2002)

Green, A., *Cultural History* (Basingstoke: Palgrave Macmillan, 2008)

Grier, D. A., *When Computers Were Human* (Princeton, NJ: Princeton University Press, 2005)

Griffen, M., *Amateur photography and pictorial aesthetics: influences of organization and industry on cultural production*, PhD thesis, University of Pennsylvania (1987)

Hamilton, E., *Crashing Suns* (New York: Ace, 1965)

Harris, T. J., United States Office of Naval Research and International Business Machines Corporation, *Holographic Head-Up Display: Phase I* (Poughkeepsie, NY: International Business Machines Corporation, 1967)

Hayward, P., *Culture, Technology and Creativity in the Late Twentieth Century* (London: J. Libbey, 1990)

Heims, S. J., *Constructing a Social Science for Postwar America: The Cybernetics Group, 1946–1953* (Cambridge, MA: MIT Press, 1993)

Hentschel, K., *Mapping the Spectrum: Techniques of Visual Representation in Research and Teaching* (Oxford: Oxford University Press, 2002)

Hentschel, K., *Visual Cultures in Science and Technology: A Comparative Perspective* (Oxford: Oxford University Press, 2014)

Hentschel, K. and A. D. Wittmann (eds.), *The Role of Visual Representations in Astronomy: History and Research Practice* (Frankfurt am Main: Verlag Harri Deutsch, 2000)

Hepworth, C. M., *Came the Dawn: Memories of a Film Pioneer* (London: Phoenix House, 1951)

Herbert, D., *Mr Wizard's Science Secrets* (USA: Popular Mechanics Co., 1952)

Hewison, R., *Too Much: Art and Society in the Sixties 1960–1975* (London: Methuen, 1986)

Hogan, M. J., *The Marshall Plan: America, Britain, and the Reconstruction of Western Europe, 1947–1952* (Cambridge: Cambridge University Press, 1987)

Hughes, R., *The Shock of the New: Art and the Century of Change* (New York: Thames and Hudson, 1991)

Hughes, T. P., *Networks of Power: Electrification in Western Society, 1880–1930* (Baltimore: Johns Hopkins University Press, 1983)

Hughes, T. P., *American Genesis: A Century of Invention and Technological Enthusiasm* (New York: Viking, 1989)

Ingalls, A. G., *Amateur Telescope Making* (New York: Scientific American, 1933 (3rd edition))

Jankovic, V., *Reading the Skies. A Cultural History of the English Weather, 1650–1820* (Manchester: Manchester University Press, 2001)

Jenkins, R. V., *Images and Enterprise: Technology and the American Photographic Industry 1839 to 1925* (Baltimore: Johns Hopkins University Press, 1975)

Jeong, T. H., *Gaertner–Jeong Holography Manual* (Chicago: Gaertner Scientific Corp., 1968)

Jeong, T. H., *A Study Guide on Holography (Draft)* (Lake Forest, IL: Lake Forest College, 1975)

Jeong, T. H., *A Demonstrated Lecture on Holography* (Chicago: Instant Replay, 1978)

Jeong, T. H. and F. E. Lodge, *Holography using a Helium–Neon Laser* (Blackwood, NJ: Metrologic Instruments, 1980)

Jeske, M. A., *A high school optics unit emphasizing laser experiments and student production of various hologram types*, thesis, (1994)

Joerges, B. and T. Shinn (eds.), *Instrumentation: Between Science, State and Industry* (Dordrecht: Kluwer Academic, 2001)

Johnson, D., *Edward Stratemeyer and the Stratemeyer Syndicate* (New York: Twayne, 1993)

Johnson, G. T., *A holography unit for a junior high electronics laboratory*, thesis, (1983)

Johnston, S. F., *A History of Light and Colour Measurement: Science in the Shadows* (Bristol: Institute of Physics Publishing, 2001)

Johnston, S. F., *Holographic Visions: A History of New Science* (Oxford: Oxford University Press, 2006)

Johnston, S. F., *The Neutron's Children: Nuclear Engineers and the Shaping of Identity* (Oxford: Oxford University Press, 2012)

Jones, C. A. and P. Galison (eds.), *Picturing Science, Producing Art* (New York: Routledge, 1998)

Jordan, F. I., *Photographic Enlarging* (Rochester, NY: Folmer Graflex Corp, 1935)

Jussim, E., *Visual Communication and the Graphic Arts: Photographic Technologies in the Nineteenth Century* (New York: R. R. Bowker, 1974)

Kamm, A. and M. Baird, *John Logie Baird: A Life* (Edinburgh: National Museums of Scotland, 2002)

Kammen, M., *Visual Shock: A History of Art Controversies in American Culture* (New York: Random House, 2006)

Keystone View Company Education Department, *Visual Education: Teacher's guide to the Keystone '600 Set'*. (Meadville, PA: Keystone View Company, 1920)

Kirk, A. G., *Counterculture Green: The Whole Earth Catalog and American Environmentalism* (Lawrence: University Press, Kansas, 2011)

Klüver, B., J. Martin and B. Rose (eds.), *Pavilion: Experiments in Art and Technology* (New York: E. P. Dutton, 1972)

Lafollette, M. C., *Science on American Television: A History* (Chicago: University of Chicago Press, 2013)

Latour, B., *Science in Action: How to Follow Scientists and Engineers through Society* (Milton Keynes: Open University Press, 1987)

Lévi-Strauss, C., *La Pensée Sauvage* (Paris: Librairie Plon, 1962)

Lissack, L. and S. Lissack, *Dali in Holographic Space* (e-book: CreateSpace, 2012)

Lloyd, V., *Photography: The First Eighty Years* (London: Colnaghi, 1976)

Lodder, C., *Russian Constructivism* (New Haven, CT: Yale University Press, 1983)

Lootens, J. G., *On Photographic Enlarging and Print Quality* (New York: The Camera, 1944)

Lund, P., *Photography as a Hobby* (Bradford: Percy Lurid, Humphries & Co, 1895)

Lynch, A., *Thought Contagion: How Belief Spreads Through Society* (New York: BasicBooks, 1996)

Maddox, T., *Snake-Eyes* (New York: Omni Publications, 1986)

Marcuse, H., *One-Dimensional Man: Studies in the Ideology of Advanced Industrial Society* (Boston: Beacon Press, 1964)

Marien, M. W., *History of Photography: A Cultural History* (London: Laurence King, 2002)

Marvin, C., *When Old Technologies Were New: Thinking About Electric Communication in the Late Nineteenth Century* (Oxford: Oxford University Press, 1988)

Mason, L. C., *All about Making Darkroom Gadgets with Your Own Hands* (London: Focal Press, 1954)

McDougall, W. A., *The Heavens and the Earth: A Political History of the Space Age* (New York: Basic Books, 1985)

McKay, H. C., *The Photographic Negative (Vols. I-IV)* (New York: Ziff-Davis, 1942)

McLuhan, M., *Understanding Media: The Extensions of Man* (New York: McGraw-Hill, 1964)

Meadows, J., *The Victorian Scientist: The Growth of a Profession* (London: British Library, 2004)

Mees, C. E. K., *From Dry Plates to Ektachrome Film: A Story of Photographic Research* (New York: Ziff-Davis, 1961)

Meggs, P. B., *A History of Graphic Design* (New York: John Wiley & Sons, 1998)

Meikle, J. L., *Plastic: A Cultural History* (New Brunswick, NJ: Rutgers University Press, 1997)

Moran, J., *Printing Presses: History and Development from the Fifteenth Century to Modern Times* (Berkeley: University of California, 1973)

Moran, J., *Armchair Nation: An Intimate History of Britain in Front of the TV* (London: Profile, 2013)

Morus, I. R., *Frankenstein's Children: Electricity, Exhibition, and Experiment in Early-Nineteenth-Century* (Princeton, NJ: Princeton University Press, 1998)

Newton, I., *Opticks, or a Treatise on the Reflexions, Refractions, Inflexions and Colours of Light* (London: Smith & Walford, 1704)

Niven, L., *Ringworld* (New York: Ballantine, 1970)

Nye, D. E., *American Technological Sublime* (Cambridge, MA: MIT Press, 1994)

Oudshoorn, N. and T. Pinch (ed.), *How Users Matter: The Co-Construction of Users and Technology* (Cambridge, MA: MIT Press, 2005)

Pepper, J. H., *The True History of the Ghost: And All About Metempsychosis* (London: Cassell, 1890)

Pethick, J., *On Holography and a Way to Make Holograms* (Burlington, ON: Belltower Enterprises, 1971)

Phillips, D., *Art for industry's sake: halftone technology, mass photography, and the social transformation of American print culture 1880–1920*, PhD thesis, Yale University (1996)

Picardo, R., *Star Trek Voyager: The Hologram's Handbook* (Pocket Books, 2002)

Reader, W. J., *Imperial Chemical Industries: A History* (Oxford: Oxford University Press, 1975)

Reich, C. A., *The Greening of America: How the Youth Revolution is Trying to Make America Livable* (London: Allan Lane, 1970)

Reich, L. K., *The Making of Industrial Research, Science and Business at GE and Bell* (New York: Cambridge University Press, 1985)

Rhees, D. J., *A new voice for science: science service under Edwin E. Slosson, 1921–1929*, MA thesis, History, University of North Carolina at Chapel Hill (1979)

Rhody, A. and F. Ross (eds.), *Holography Marketplace 7th Edition* (Berkeley, CA: Ross Books, 1998)

Rhody, A. and F. Ross, *Holography Marketplace: The Reference Text and Sourcebook for Holography Worldwide* (Berkeley, CA: Ross Books, 1999)

Robbins, L. K., *Holography, a new medium for the high school art curriculum*, MA thesis, Ball State University (1980)

Robinson, H. P., *Pictorial Effect in Photography: Being Hints on Composition and Chiaroscuro for Photographers* (London: Piper & Carter, 1869)

Roszak, T., *The Making of a Counter-Culture: Reflections on Technocratic Society and Its Youthful Opposition* (London: Faber and Faber, 1970)

Royle, N., *The Uncanny* (Manchester: Manchester University Press, 2003)

Rudolphs, J. L., *Scientists in the Classroom: The Cold War Reconstruction of American Science Education* (New York: Palgrave, 2002)

Sabra, A. I., *The Optics of Ibn al-Haytham (Books I-V)* (Kuwait: National Council for Culture, Arts and Letters, 1983–2002)

Sammons, E., *The World of 3-D Movies* (Valencia: Delphi, 1992)

Saunders, F. S., *The Cultural Cold War: The CIA and the World of Arts and Letters* (New York: W. W. Norton, 1999)

Saxby, G., *Practical Holography* (New York: Prentice-Hall International, 1987)

Schroeter, J., *3D: History, Theory and Aesthetics of the Transplane Image* (New York: Bloomsbury Academic, 2014)

Schumacher, E. F., *Small is Beautiful: Economics as if People Mattered* (New York: Harper & Row, 1973)

Seiberling, G. and C. Bloore, *Amateurs, Photography, and the Mid-Victorian Imagination* (Chicago: University of Chicago Press, 1986)

Shapin, S., *The Scientific Life: A Moral History of a Late Modern Vocation* (Chicago: University of Chicago Press, 2009)

Shifman, L., *Memes in Digital Culture* (Cambridge, MA: MIT Press, 2013)
Sidgwick, J. B., *Amateur Astronomer's Handbook* (London: Faber & Faber, 1955)
Stearns, P. N., *Consumerism in World History* (New York: Routledge, 2001)
Stebbins, R. A., *Amateurs, Professionals and Serious Leisure* (Montreal: McGill University Press, 1992)
Stong, C. L. (ed.), *The Scientific American Book of Projects for the Amateur Scientist* (New York: Simon & Schuster, 1960)
Stowman, W. P., *A holography laboratory for high school physics*, thesis, (1972)
Sturken, M. and L. Cartwright, *Practices of Looking: An Introduction to Visual Culture* (Oxford: Oxford University Press, 2009)
Terzian, S. G., *Science Education and Citizenship: Fairs, Clubs, and Talent Searches for American Youth, 1918–1958* (New York: Palgrave Macmillan, 2013)
Thurlow, C., *Sex, Surrealism, Dalí and Me: The Memoirs of Carlos Lozano* (Cornwall: Razor Books, 2000)
Toffler, A., *Future Shock* (London: Pan Books, 1971)
Toffler, A., *The Third Wave* (New York: Morrow, 1980)
Tupitsyn, M., *The Soviet Photograph, 1924–1937* (New Haven, CT: Yale University Press, 1996)
Turner, F., *From Counterculture to Cyberculture: Stewart Brand, the Whole Earth Network, and the Rise of Digital Utopianism* (Chicago: University of Chicago, 2006)
Unterseher, F., J. Hansen and B. Schlesinger, *Holography Handbook: Making Holograms the Easy Way* (Berkeley, CA: Ross Books, 1981)
Varley, J., *The Ophiuchi Hotline* (New York: Dial Press, 1977)
Walker, J. A. and S. Chaplin, *Visual Culture: An Introduction* (Manchester: Manchester University Press, 1997)
Wang, J., *American Science in an Age of Anxiety: Scientists, Anticommunism, and the Cold War* (Chapel Hill: University of North Carolina Press, 1999)
Ware, C. F. (ed.), *The Cultural Approach to History* (New York: Columbia University Press, 1940)
Warhol, A., *The Philosophy of Andy Warhol (From A to B and Back Again)* (San Diego, CA: Harcourt, 1975)
Wasserman, S. and K. Faust, *Social Network Analysis: Methods and Applications* (Cambridge: Cambridge University Press, 1994)
Weinberg, A. M., *Reflections on Big Science* (Boston: MIT Press, 1967)
Wiener, N., *Cybernetics and Society: The Human Use of Human Beings* (Boston: Houghton Mifflin, 1954)
Wilber, K., *The Holographic Paradigm and Other Paradoxes: Exploring the Leading Edge of Science* (Boulder, CO: Shambhala, 1982)
Wilson, R. G., D. H. Pilgrim and D. Tashjian, *The Machine Age in America: 1918–1941* (New York: Brooklyn Museum, 1986)
Winner, L., *Autonomous Technology: Technics-out-of-Control as a Theme in Political Thought* (Cambridge, MA: MIT Press, 1978)
Yaroslavsky, L. P. and N. S. Merzlyakov, *Methods of Digital Holography* (New York: Consultants Bureau, 1980)
Zone, R., *Stereoscopic Cinema and the Origins of 3-D Film, 1838–1952* (Lexington, KY: University Press of Kentucky, 2007)

PAPERS, ARTICLES AND OTHER MEDIA

'2331. Great Military Review on Emperor's Birthday, His Majesty attending, Qoyama Parade Grounds, Tokyo, Japan: stereo view', (H. C. White Co, 1905)
'3-D lasography – the month old giant', *Laser Focus*, 1 Jan 1965: 10–15

'American Scenery, New England Series: stereo view', (c1875)
'Atari Cosmos game system', *Holosphere* 10 (1981): 1
'Do-it-yourself holography', *Rolling Stone*, (18), 30 Aug 1973: 40
'Editorial', *British Journal of Photography*, 38, 20 Feb 1891: 17
'Entrepreneurs urged to set up holographic centers', *Holosphere* 5 (6) (1976): 4, 8
'Going to church in Sweden: stereo view', (Pettijohn Packages, 1906)
'Holograms as giftware: the commercialization of artistic holography', *Holography Marketplace* 8 (1999): 37–43
'Holographic Image: a Virginia-based company is keeping busy with the growing popularity of holographic materials because it supplies the embossed materials as well as the application equipment', *Paper, Film and Foil Converter* 68 (1994): 90
'Holographic playmates', *Holography News*, 15 (2), Mar 2001: 4
'Holographic storage promises high data density', *Laser Focus World* 32 (1996): 81–2, 84–5, 87–8, 90, 93
'Holographic tv: picture is bright', *Electronics*, 28 Nov 1966: 25
'Holography – photography in three dimensions at Indiana Institute of Technology', *The Announcer*, 7 (2), Oct 1966: 1
Holography in the USSR: the Art and Science of the Soviet Union exhibition catalogue (York, UK: 1981)
'Holography: clearing the image', *Time*, 91 (12), 22 Mar 1968: 69
'Holotron licensing outlined by Cravens', *Holosphere* 7 (9) (1978): 3, 6
'IBM supermarket scanner uses holographic optics', *Holosphere* 9 (2) (1980): 1–2
'Into 1994: new developments in the marketplace', *Holography News*, 7 (10), Jul 1993: 88
'It's Spring, and holo galleries are popping up from coast to coast', *Holosphere*, 8 (3), Mar 1979: 1
'Laser beam creates new A-V medium', *Sales Meetings*, 15 May 1966: 52–7
'Lensless laser photography', *Popular Electronics*, 20 (3), Mar 1964: 60
'Lensless photography uses laser beams to enlarge negatives, microscope slides', *Wall Street Journal*, 5 Dec 1963, 28
'Museums and scientific progress', *Discovery*, 8 (8), Aug 1947: 228–9
'National Electronics Conference, Chicago', *Pontiac Press*, 28 Oct 1965,
'New camera operating without lens shows scientific promise', *New York Times International Edition*, 11 Dec 1963, 14
'New microscope limns molecule: Britons impressed by paper combining optical principle with electron method', *New York Times*, 15 Sep 1948, 35
'NEWS – News Briefs – Breakthrough Promises Cheaper, Faster Chips – Optical Switches Turn to Holography', *Computer* 34(2001): 3
'No-lens picture-taking devised by U-M men', *Ann Arbor News*, 5 Dec 1963, 31
'Now comes the natural vision movie – which looks like the pictures you used to see thru the stereoscope', *Motion Picture Magazine*, 33 (1), Feb 1927: 57
'Number of galleries showing holograms continues to grow', *Holosphere* 6 (1) (1977): 1
'Obituary notice: William de Wiveleslie Abney', *Proceedings of the Royal Society* A99 (1921): i–v
'Outlook – Holography and "spintronics" promise more data density I/O chip architecture enables increased bandwidth', *Electronic Products* 44 (2001): 6
'Packaging and design sells: Holography News interviews Barc Thompson', *Holography News*, 7 (1), 1993: 3–4
'Photo shows work of new optic system: Viewers call exhibit startling, fantastic', *Chicago Tribune*, 27 Oct 1965, 36

'Production equipment: dot matrix hologram machines', *Holography Marketplace* 8 (1999): 105–10
'Shoppers get the picture: hologram marketers test the water', *Union*, 20 Jan 1977, 2
'Soviet holography conference shows high amount of research activity', *Holosphere*, 7 (9), Sep 1978: 1–2
'Stereophotography sweeps the country', *Popular Mechanics*, 98 (3), Sep 1952: 106–11
'Strick will make laser film', *Los Angeles Times*, 21 Aug 1970, 14
'Stunts with your camera', *Popular Mechanics*, 59 (6), Jun 1933: 874–8
'Taking unusual photos with pinhole cameras', *Popular Mechanics*, 60 (3), Sep 1933: 428–9
'The "death ray" rivals', *New York Times*, 29 May 1924, 1
'The first hologram ever distributed to a mass audience', *Journal of the SMPTE* 76(1967): 707
'The news of radio', *New York Times*, 1 Jul 1948, 46
'The stereoscope in the schoolroom', *Pennsylvania School Journal* 17(1868): 150–2
'The world under a lens', *Popular Mechanics*, 60 (2), Aug 1933: 202–5
'Theatre technology of the future' *Theatre Design and Technology*, 15 (3), Fall 1979: special issue
3D Realms, *Duke Nukem 3D* videogame (3D Realms, 1996)
Abbe, E., 'A contribution to the theory of the microscope and the nature of microscopic vision', *Proceedings of the Bristol Naturalists' Society* 1 (1874): 200–61
ABC Studios, 'Eternity', *Perception*, series 3, episode 5 (TNT, 2014)
Albright, T., 'School of holography', *Radical Software* 2 (1972): 56–7
Allibone, T. E., 'White and black elephants at Aldermaston', *Journal of Electronics and Control* 4 (1958): 179–92
Allibone, T. E., 'Dennis Gabor 1900–1979', *Biographical Memoirs of Fellows of the Royal Society* 26 (1980): 107–40
American Institute of Physics, 'Press release: Objects behind others now visible in 3-D pictures made by new method,' New York, 1964
Arctander, E. H., 'Signs you'll see appearing out of thin air', *Popular Science*, 195 (2), Aug 1969: 81–2, 193
Arkin, W. M., 'When seeing and hearing isn't believing', *Washington Post*, 1 Feb 1999, http://www.washingtonpost.com/wp-srv/national/dotmil/arkin020199.htm, accessed 11 Dec 2014
Artner, A. G., 'Rhetoric, not results, at holography show', *Chicago Tribune*, 12 Jun 1977, 37
Austin, D., 'Exhibition: Light Fantastic 2', *New Scientist*, 19 Jan 1978: 171
Baez, A. V., 'Pinhole-camera experiment for the introductory physics course', *American Journal of Physics* 25 (1957): 636–8
Baez, A. V., 'Some notes on laboratory procedures in a small college', *American Journal of Physics* 25 (1957): 386–7
Baez, A. V., 'Anecdotes about the early days of x-ray optics', *Journal of X-Ray Science and Technology* 7 (1997): 90–7
Baez, A. V., 'The early days of x-ray optics – x-ray microscopes, telescopes and holograms,' invited talk, Lawrence Livermore National Laboratory, CA, 1999
Baez, A. V. and G. Castro, 'A laboratory demonstration of the three-dimensional nature of in-line holography', *American Journal of Physics* 67 (1999): 876–9
Bains, S., 'Editorial', *Holographics International* 1 (1989): 4
Bak, M. A., 'Democracy and discipline: Object lessons and the stereoscope in American education, 1870–1920', *Early Popular Visual Culture* 10 (2012): 147–67
Barilleaux, R. P., 'Holography and the Art World', *Leonardo* 25 (1992): 417–8
Barker, D., 'Laser affair', *Guardian*, 4 Mar 1978, 15

Bartolini, R., W. Hannan, D. Karlsons and M. Lurie, 'Embossed hologram motion pictures for television playback', *Applied Optics* 9 (1970): 2283–90

Bartolini, R. A., 'Holographic motion pictures for TV playback', presented at *1973 IEEE/OSA Conference on Laser Engineering and Applications Digest of Technical Papers*, New York, 1973

Bartolini, R. A., J. Bordogna and D. Karlsons, 'Recording considerations for RCA Holotape', *RCA Review* 33 (1972): 170–205

Bauer, R., 'Holographie – Eine kommerziell erfolgreiche aber als Kunstform gescheiterte Technologie?', in: R. Bauer, J. Williams and W. Weber (eds.), *Technik zwischen Artes und Arts [Technology between artes und arts]*. (Münster: Waxman, 2008), pp 133–47

Begleiter, E., 'Edible holography: the application of holographic techniques to food processing', presented at *Practical Holography V*, San Jose, CA, 1991

Benjamin, W. and H. Zohn (translator), 'On some motifs in Baudelaire', in: H. Arendt (ed.), *Illuminations: Essays and Reflections* (New York: Schocken, 1968), pp 155–200

Benrey, R. M., 'Make holograms at home – at a budget price', *Popular Science*, 198 (1), Jan 1971: 84–6

Benton, S. A., 'Rainbow holograms', *Journal of the Optical Society of America* 50 (1969): 1545–6

Benton, S. A., 'Achromatic images from white light transmission holograms', *Journal of the Optical Society of America* 68 (1978): 1441

Benton, S. A., S. M. Birner and A. Shirakura, 'Edge-lit rainbow holograms', *Proceedings of SPIE* 1212 (1990): 149–57

Benton, S. A., H. S. Mingace, Jr. and W. R. Walter, 'One-step white-light transmission holography', in: P. N. Tamura and T. C. Lee (eds.), *Recent Advances in Holography* (Bellingham, WA: SPIE, 1980), pp 156–61

Benyon, M., 'Pulsed holographic art practice', presented at *Practical Holography*, Los Angeles, CA, 1986

Berry, J. R., 'Holography: 3D magic in mid-air', *Popular Mechanics* 129 (1968): 104–7, 206

Bjelkhagen, H. I., 'Holographic portraits made by pulse lasers', *Leonardo* 25 (1992): 443–8

Bjelkhagen, H. I., 'Lippmann photography: its history and recent development', *The PhotoHistorian* 141–2 (2003): 11–9

Bjelkhagen, H. I. and P. G. Crosby, 'Color holography and its use in display and archiving applications', *Archiving 2008 Final Program and Proceedings* (2008): 239–45

Bjelkhagen, H. I., R. E. Deem, J. Landry, M. E. Marhic and F. D. Unterseher, 'Holographic portrait of Ronald Reagan', presented at *International Symposium on Display Holography*, Lake Forest, IL, 1991

Bjelkhagen, H. I. and D. Vukicevic, 'Color holography: a new technique for reproduction of paintings', *Proceedings of SPIE* 4659 (2002): 83–90

Bleiler, E. F., 'From the Newark Steam Man to Tom Swift', *Extrapolation* 30 (2) (1989): 101–16

Bohm, D., 'Quantum theory as an indication of a new order in physics. II. Implicate and explicate order in physical law', *Foundations of Physics* 3 (1973): 139–68

Bove Jr., V. M., 'Display holography's digital second act', *Proceedings of the IEEE* 100 (2012): 918–27

Bowen, M., 'Laser days', *Guardian*, 12 Jan 1978, 8

Boyle, D., 'The flying hologram: latest in head-up display systems', *Interavia (English Edition)* 36 (1981): 44–5

Brenden, B. B., 'Acoustical holography – some techniques of promise', *Materials Evaluation* 31 (1973): 40A

Brown, E. H., 'Rationalizing consumption: Lejaren á Hiller and the origins of American advertising photography, 1913–1924', *Enterprise and Society* 1 (2000): 715–38

Brown, B. R. and A. W. Lohmann, 'Computer-generated binary holograms', *IBM Journal of Research and Development* 13 (1969): 160–8

Bryant, E., 'The poet in the hologram in the middle of prime time', in: H. Harrison (ed.), *Nova 2: Original Science Fiction Stories* (London: Sphere Books, 1972), pp 116–32

Buchli, V., 'Introduction', in: V. Buchli (ed.), *The Material Culture Reader* (Oxford: Oxford University Press, 2002), pp 1–22

Bungie, *Halo* videogame (Microsoft, 2000)

Bureau of Standards, 'Construction and operation of a very simple radio receiving equipment', Letter Circular LC 43, 16 Mar 1922

Burns Jr, J., 'Update on the New York School of Holography', *Holosphere*, 4 (4), April 1975: 3–4

Bush, E. A., 'Hologram cylinders, 4 meters long, display rare objects without risk', *Holosphere*, 7 (10), Oct 1978: 1–2

Bush, V., 'Science The Endless Frontier: A Report to the President by Vannevar Bush, Director of the Office of Scientific Research and Development', United States Printing Office, July 1945

Cato, R. T., 'Hand-held holographic scanner having highly visible locator beam', *IBM Technical Disclosure Bulletin* 27 (1984): 2021–2

Charnetski, C., 'The impact of holography on the consumer', presented at *Electronics and Aerospace Systems Conference Convention*, Washington DC, 1970

Christy, E. H. and K. K. Sutherlin, 'Validating credit cards using holography', *IEEE International Convention Digest* 4 (1970): 338–9

Clarke, A. C., 'Hazards of prophecy: the failure of imagination', in: A. C. Clarke (ed.), *Profiles of the Future: An Enquiry into the Limits of the Possible* (London: The Scientific Book Club, 1962), p. 36

Clark, A., 'Making an impact: holographic applications for print, and looks at developments in foiling and embossing', *The British Printer* 109 (1996): 34

Cochran, G. D. and R. D. Buzzard, 'The new art of holography', in: *Science Year: The World Book Science Annual 1967* (Chicago: Field Enterprises Educational Corp, 1967), pp 200–11

Coffey, D. F. W., Pat. No. US 2063985 A 'Apparatus for making a composite stereograph' (1936), assigned to Winnik Stereoscopic Processes Inc

Connes, P., 'Silver salts and standing waves: the history of interference colour photography', *Journal of Optics* 18 (1987): 147–66

Connors, B. A., 'Lightforest and the MIT Museum Holography Education Project', presented at *Sixth International Symposium on Display Holography*, Lake Forest, IL, 1997

Cordes, C., 'Bruce Nauman: Prints 1970–89,' New York, 1989

Cowan, R. S., 'The consumption junction: a proposal for research strategies in the sociology of technology', in: W. E. Bijker, T. P. Hughes and T. J. Pinch (eds.), *The Social Construction of Technological Systems* (Cambridge, MA: MIT Press, 1987), pp 261–80

Cowles, S., 'The New York experience', *Holographics International*, 1 (2), Winter 1988: 14–5

Cranbrook Academy of Art, 'The Laser: Visual Applications,' Detroit, MI, 1969

Cross, L., 'The potential impact of the laser on the video medium', *Radical Software* 1 (1970): 6

Cross, L. G. and C. Cross, 'HoloStories: reminiscences and a prognostication on holography', *Leonardo* 25 (1992): 421–4

Cross, L. and J. Pethick, 'Lloyd Cross and Gerald Pethick on holography', *Ode to Gravity* (KFPA, 20 Oct 1971)

Crowther, B., 'This is Cinerama (1952): new movie projection shown here; giant wide angle screen utilized; novel technique in films unveiled', *New York Times*, 1 Oct 1952, 1, 40

Crytek, *Crysis 2* videogame (Electronic Arts, 2011)

Davis, W., '*Adventures in Science*: Fourth Annual Science Talent Search', CBS radio script, 17 Feb 1945, SIA RU7091, Box 391, folder 1

Davy, J., '3-D pictures in the round', *Observer*, 15 May 1966, 13

de Closets, F., 'Le laser «voit» l'invisible: Révolution en optique: le relief total en photographie', *Sciences et Avenir* (1964): 162–7, 212

de Plante, R., 'How computers clear up fuzzy pictures', *Deseret News (Arizona)*, 24 Sep 1969, A11

del Sesto, S. L., 'Wasn't the future of nuclear energy wonderful?', in: J. J. Corn (ed.), *Imagining Tomorrow: History, Technology, and the American Future* (Cambridge, MA: MIT Press, 1986), pp 58–76

Denisyuk, Y. N., 'On the reflection of optical properties of an object in the wave field of light scattered by it', *Doklady Akademii Nauk SSSR* 144 (1962): 1275–8

Denisyuk, Y. N., 'On the problem of a photograph reproducing the full illusion of the reality of the object depicted (in Russian)', *Zhurnal Nauchnoi i Prikladnoi Fotografi i Kinematografi* 11 (1966): 46–56

Denisyuk, Y. N., 'Holography', *Soviet Union* 9 (1970): 12–4

Denisyuk, Y. N., 'Holography at the State Optical Institute (GOI)', *Soviet Journal of Optical Technology* 56 (1989): 38–43 Translated from *Optiko Mekhanicheskaya Promyshlennost* 56 (1989): 40–4

Denisyuk, Y. N., 'My way in holography', *Leonardo* 25 (1992): 425–30

Denisyuk Yu, N., 'Imaging properties of light intensity waves: the development of the initial Lippmann ideas', *Journal of Optics* 22 (1991): 275–80

Deschin, J., 'No-lens pictures: photographic technique employs light alone', *New York Times*, 15 Dec 1963, 143

Dickson, L. D. and C. L. Stong, 'The Amateur Scientist: stability of the apparatus: insuring a good hologram by controlling vibration and exposure', *Scientific American*, 224 (7), July 1971: 110–7

Dodd, P., 'A spy in the sky! Army unveils radar camera: it shoots enemy thru clouds of smoke', *Chicago Daily Tribune*, 20 Apr 1960, 3

du Maurier, G., 'Cartoon', *Punch's Almanac*, 8 Dec 1879:

Edgerton, D. E. H., 'Industrial research in the British photographic industry, 1879–1939', in : J. Liebenau (ed.), *The Challenge of New Technology* (Aldershot: 1988), pp 106–34

Editor's Choice, 'Holography ideas', *Potentials in Marketing*, Sep 1991: 14

Edson, L., 'Box cameras were never like this', *Popular Mechanics* 126 (Jul 1966): 67–9, 171, 3

Edwards, R., 'The "citizens" in citizen science projects: educational and conceptual issues', *International Journal of Science Education, Part B* 4 (2014): 376–91

Eisenhower, D. D., 'Farewell address to the nation', 17 Jan 1961

Fairstein, J., 'The San Francisco School of Holography', http://www.jfairstein.com/SOH.html, accessed 28 Feb 2003

Feaver, W., 'Crafty stuff: art', *Observer*, 16 Dec 1979, 18

Finch College. Museum of Art. Contemporary Study Wing., T. McBurnett and E. H. Varian, 'N Dimensional Space,' New York, 1970

Forster, H. D., Pat. No. US 3844645 A 'Three-dimensional projecting system' (1974), assigned to Holograph Corporation

Gabor, D., Pat. No. 685,286 'Improvements in and relating to Microscopy' (1947), assigned to British Thomson-Houston

Gabor, D., 'A new microscopic principle', *Nature* 161 (1948): 777–8

Gabor, D., 'Holography, 1948–1971', in: Nobel Prize Committee (ed.), *Les Prix Nobel En 1971* (Stockholm: Imprimerie Royale P. A. Norstedt & Sonner, 1971), pp 169–201

Geelen, P., 'Holography in education: it's time to come out of the dark', *The Australian Science Teachers' Journal* 40 (1994): 15

Geoghegan, H., 'Emotional geographies of enthusiasm: belonging to the Telecommunications Heritage Group', *Area* 45 (2012): 40–6

Gibson, W., 'The Gernsback continuum', in: W. Gibson (ed.), *Burning Chrome* (New York: Arbor House, 1986), pp 24–36

Gibson, W., 'Fragments of a hologram rose', in: W. Gibson (ed.), *Burning Chrome* (New York: Ace Books, 1987), pp 31–9

Gilman, L. C., 'The promise of holography', *Bell Telephones Magazine* 46 (1967): 15–8

Gilmore, C. P., 'Those incredible holograms: amazing "frozen light waves" give first true 3D pictures', *Popular Science* 189(1966): 78–91, 174–5

Gilmore, C. P., 'How lasers are going to work for you', *Popular Science* 197 (1970): 48–53

Glueck, G., 'Works made with aid of laser on display at Finch: Art holograms in their infancy', *New York Times*, 25 Apr 1970, 24

Good, J. P., 'The scope and outlook of visual education', *School Science and Mathematics* 20 (1920): 482–7

Gorglione, N., 'Lloyd Cross', *Holographics International* 1 (1987): 17, 29

Gorglione, N., 'Forms of light: a personal history in holography', *Leonardo* 25 (1992): 473–80

Gosling, N., 'Beaming pioneer', *Observer*, 1 Mar 1970, 33

Griffen, M., 'Between art and industry: amateur photography and middle-brow culture', in: L. Gross (ed.), *On the Margins of Art Worlds* (Boulder, CO: Westview Press, 1995), pp 183–205

Groskinski, H., 'Cassette TV: the good revolution', *Life*, 69 (16), 16 Oct 1970: 46–53

Gruen, J., 'The whole message', *New York*, 3 (19), 11 May 1970: 59

Gunning, T., 'The cinema of attraction: early film, its spectator, and the Avant-Garde', *Wide Angle* 8 (1986): 63–70

Gunning, T., 'An aesthetic of astonishment: the (in)credulous spectator', *Art and Text* 34 (1989): 31–45

Gunning, T., 'Hand and eye: excavating a new technology of the image in the Victorian era', *Victorian Studies* 54 (2012): 495–516

Hagen, C., 'The case for holograms: the defense resumes', *New York Times*, 29 Nov 1991, C24

Haines, K., 'Development of embossed holograms', *Proceedings of SPIE* 2652 (1996): 45–52

Haley, C., 'Science fiction and the making of the laser', in: M. Higgins, G. Lightfoot, M. Parker and W. Smith (eds.), *Science Fiction and Organization* (New York: Routledge, 2001), pp 31–9

Halle, M. W., S. A. Benton, M. A. Klug and J. S. Underkoffler, 'Ultragram: a generalized holographic stereogram', *Practical Holography V* 1461 (1991): 132–41

Goldfinger, Dir. Hamilton, G., United Artists (1964)

Harrison, G. B., 'The laboratories of Ilford Limited', *Proceedings of the Royal Society of London Series B, Biological Sciences* 142 (1954): 9–20

Harvey, T. W., 'Holotron reports results of holography market survey', *Holosphere*, 4 (4), April 1975: 1–3

Hasbro, *Jem and the Holograms* (Claster Television, 1995–2001)

Havighurst, R. J. and K. Feigenbaum, 'Leisure and lifestyle', *American Journal of Sociology* 64 (1959): 396–404

Hecht, J., 'Applications pioneer interview: Emmett Leith', *Lasers & Applications*, 5, April 1986: 56–8

Hector, R., 'Interview with Stilphen', http://www.2600connection.com/interviews/roger_hector/interview_roger_hector.html, accessed 15 Aug 2014

Hess, T. B., 'Hocus-pocus and hocus-focus', *New York*, 8 (33), 18 Aug 1975: 70–1

Heumann, S. M. and C. L. Stong, 'The Amateur Scientist: How to make holograms and experiment with them or with ready-made holograms', *Scientific American*, 216 (2), Feb 1967: 122–8

Hoffman, E. P., 'Soviet views of the "Scientific-Technical Revolution"', *World Politics* 30 (1978): 615–44

Holmes, O. W., 'The stereoscope and the stereograph', *The Atlantic Monthly* III (1859): 738–48

Homburg, E., 'The emergence of research laboratories in the dyestuffs industry 1870–1900', *British Journal for the History of Science* 25 (1992): 91–111

Hughes, T. P., 'The seamless web: technology, science, etcetera, etcetera', *Social Studies of Science* 16 (1986): 281–92.

Hunt, G. C., 'The impact of visual communications on industry', *Business Screen* 28 (1967): 24–5

Hurter, F. and V. C. Driffield, 'Photochemical investigations and a new method of determination of the sensitiveness of photographic plates', *Journal of the Society of Chemical Industry* 9 (1890): 455–69

Iizuka, K. and C. L. Stong, 'The Amateur Scientist: holograms with sound and radio waves', *Scientific American*, 226 (11), Nov 1972: 120

Ikeda, H., S. Matsumoto and T. Inagaki, 'Hologram scanner for POS bar code symbol reader', *Fujitsu Scientific and Technical Journal* 15 (1979): 59–76

Insomniac Games, *Ratchet & Clank* videogame (Sony, 2002)

Jackson, R., Through the Looking Glass: An exhibition of three-dimensional, floating images made with lasers (Oregon Museum of Science and Industry, 1980)

James, R., 'Off the wall', *Holosphere*, 8 (12), Dec 1979: 3

Jenkins, R. V., 'Technology and the market: George Eastman and the origins of mass amateur photography', *Technology and Culture* 16 (1975): 1–19

Jensen, H., L. C. Graham, L. J. Porcello and E. N. Leith, 'Side-looking airborne radar', *Scientific American*, 237 (10), Oct 1977: 84–95

Jeong, T. H., 'Holography comes out of the cellar', *Optical Spectra*, 2 (6), Jun 1968: 58–65

Jeong, T. H., 'Holography and education in the United States', presented at *Three-Dimensional Holography: Science, Culture, Education*, Kiev, USSR, 1989

Jeong, T. H., 'Making holograms in middle and high schools', in *Proceedings of the SPIE Sixth International Conference on Education and Training in Optics and Photonics*, Cancun, Mexico, 2000

Jeong, T. H. and V. B. Markov, Three-dimensional holography—science, culture, education: the international UNESCO seminar: 5–8 September 1989, Kiev, USSR (Bellingham, WA: SPIE, 1991)

Jeong, T. H., P. Rudolf and J. Luckett, '123 360 degree holography', *Journal of the Optical Society of America* 56 (1966): 1263

Jet Propulsion Laboratory Office of Public Information, 'JPL computer process brightens Surveyor moon pictures', press release, 9 Aug 1966, http://www.jpl.nasa.gov/news/releases/60s/release_1966_0402.html

John, P., 'Teaching holography – inspiring an interest in science', *The Holographer*, http://holographer.org/media/articles/hg00003.pdf, 11 Apr 2004

Johnson, C., '*Bricoleur* and *bricolage*: from metaphor to universal concept', *Paragraph* 35 (2012): 355–72

Johnston, S. F., 'In search of space: Fourier spectroscopy 1950–1970', in: B. Joerges and T. Shinn (eds.), *Instrumentation: Between Science, State and Industry* (Dordrecht: Kluwer Academic, 2001), pp 121–41

Johnston, S. F., 'Telling tales: George Stroke and the historiography of holography', *History and Technology* 20 (2004): 29–51

Johnston, S. F., 'Attributing scientific and technical progress: the case of holography', *History and Technology* 21 (2005): 367–92

Johnston, S. F., 'From white elephant to Nobel Prize: Dennis Gabor's wavefront reconstruction', *Historical Studies in the Physical and Biological Sciences* 36 (2005): 35–70

Johnston, S. F., 'Absorbing new subjects: holography as an analog of photography', *Physics in Perspective* 8 (2006): 164–88

Johnston, S. F., 'Der parallaktische Blick: Der militärische Ursprung der Holographie [The parallax view: the military foundations of holography]', in: S. Rieger and J. Schröter (eds.), *Das holografische Wissen* (Dortmund, Germany: Diaphane, 2009), pp 33–57

Johnston, S. F., 'Why display? Representing holography in museum collections', in: P. Morris and K. Staubermann (eds.), *Illuminating Instruments* (London: Smithsonian Museum Scholarly Press, 2010), pp 97–116

Johnston, S. F., 'Security and the shaping of identity for nuclear specialists', *History and Technology* 27 (2011): 123–53

Johnston, S. F. and C. L. Stong, 'The Amateur Scientist: a high school student builds a recording spectrophotometer', *Scientific American*, 231 (1), Jan 1975: 118–25

Kakichashvili, Sh. D., 'Reconstruction of focused hologram in white light', *Ukrayins'kyi Fizychnyi Zhurnal* 14 (1969): 1862–6

Kay, W., 'Why laser men are beaming', *Sunday Times*, 17 Apr 1983, 54

Keystone View Company, p. a., 'Oliver Wendell Holmes was right: the stereoscope is not a toy', *The Educational Screen* 16 (1937): 129

Keystone View Company, p. a., 'The stereoscope goes to war', *The Educational Screen* 16 (1942): 275

Klug, M. A., M. W. Halle and P. M. Hubel, 'Full color Ultragrams', *Proceedings of SPIE* 1667 (1992): 110–9

Kontnik, L. T., 'Governments underwrite holography industry', *Holography News*, 1 (4), 27 Dec 1987: 1

Kontnik, L. T., 'Holography industry market survey and industry report: 1997', presented at *Sixth International Symposium on Display Holography*, Lake Forest, IL, 1997

Kramer, H., 'Holography: a technical stunt', *New York Times*, 20 Jul 1975, D1–D2

Kunkle, G. C., 'Technology in the seamless web: "success" and "failure" in the history of the electron microscope', *Technology and Culture* 36 (1995): 80–103

LaMacchia, J. T., 'Holographic optical memories: promise and problems', in: B. J. Thompson and J. B. Develis (eds.), *Developments in Holography II* (Bellingham WA: SPIE, 1971), pp 51–3

Lancaster, I. M., 'The Museum of Holography – its role and policies in a changing environment', presented at *Practical Holography II*, Los Angeles, CA, 1987

Lancaster, I. M. and L. T. Kontnik, 'Market conditions for display holography in the USA and Europe', presented at *Practical Holography IV*, Los Angeles, CA, 1990

Lancaster, I. M., 'The future for security applications of optical holography', *Proceedings of SPIE* 2577 (1995): 71–6

Lancaster, I. M. and L. T. Kontnik, 'International hologram manufacturers Association (IHMA): hologram counterfeiting and the hologram image register', presented at *Fifth International Symposium on Display Holography*, Lake Forest, IL, 1994

Lankford, J., 'Amateurs versus professionals: the controversy over telescope size in late Victorian science', *Isis* 72 (1981): 11–28

Larkin, A. I., 'Holography and education in the USSR', *International Symposium on Display Holography* 1600 (1991): 412–7

Latham, R. E., 'Holography in the science classroom', *Physics Teacher* 24 (1986): 395–400

Lauk, M., 'Holography in museums —a modern Sleeping Beauty syndrome', presented at *Holography in the Modern Museum*, Leicester, UK, 2008

Lawrence, P. J., 'Retailing holography in the U.S.', *Holosphere*, 13 (3), Summer 1985: 8–9

Lecht, C. P., 'A man's hologram is his castle: emergence of artificial experience', *Computerworld*, 17 (14), 4 Apr 1983: 47–50

Leith, E. N., 'Proposal for Applications of the Wavefront Reconstruction Technique', Grant proposal ORA-63–1255-PB1, Institute of Science and Technology, University of Michigan to US Army Research Office, Durham, NC, 28 May 1963

Leith, E. N., 'Electro-optics and how it grew in Ann Arbor', *Optical Spectra*, 4 (5), May 1971: 25–6

Leith, E. N., 'The legacy of Dennis Gabor', *Optical Engineering* 19 (1980): 633–5

Leith, E. N., 'Holography – the promise fulfilled', *Proceedings of SPIE* 532 (1985): 2–5

Leith, E. N., 'Reflections on display holography', *L.A.S.E.R. News*, 10 (4), 1990: 1

Leith, E. N. and J. Upatnieks, 'New techniques in wavefront reconstruction', *Journal of the Optical Society of America* 51 (1961): 1469

Leith, E. N. and J. Upatnieks, 'Reconstructed wavefronts and communication theory', *Journal of the Optical Society of America* 52 (1962): 1123–30

Leith, E. N. and J. Upatnieks, 'Wavefront reconstruction with continuous-tone transparencies', *Journal of the Optical Society of America* 53 (1963): 522

Leith, E. N. and J. Upatnieks, 'Wavefront reconstruction with diffused illumination and three-dimensional objects', *Journal of the Optical Society of America* 54 (1964): 1295–302

Leith, E. N. and J. Upatnieks, 'Photography by laser', *Scientific American*, 212 (6), Jun 1965: 24–36

Levatter, J. and C. L. Stong, 'The Amateur Scientist: a carbon dioxide laser is constructed by a high school student in California', *Scientific American*, 225 (3), Sep 1971: 218–24

Li, Y., T. Wang, S. Yang, S. Zhang, S. Fan and H. Wen, 'Theoretical and experimental study of dot matrix hologram', *Proceedings of SPIE* 3559 (1998): 121–9

Lightfoot, D. T., 'Contemporary art-world bias in regard to display holography: New York City', *Leonardo* 22 (1989): 419–24

Lindgren, N., 'Search for a holography market', *Innovations* (1969): 16–27

Lippmann, G., 'La photographie des couleurs', *Comptes Rendus de L'Academie des Sciences* 112 (1891): 274

Lippmann, G., 'Sur la théorie de la photographie des couleurs simples et composés', *Journal de Physique* 3-e serie 3 (1894): 97–107

Lippmann, G., 'Colour photography', *1908 Nobel Prize Lecture* (1908)

Lippmann, G., 'Épreuves réversibles. Photographies intégrales', *Comptes Rendus de l'Académie des Sciences* 146 (1908): 446–51

Lloyd, S., 'The Holography Unit at the Royal College of Art', *Holosphere*, 14 (3), Summer 1986: 15–6

Lohmann, A. W., 'Optical single side band transmission applied to the Gabor microscope', *Optica Acta* 3 (1956): 97–9

Lucaites, J. L. and R. Hariman, 'Visual rhetoric, photojournalism, and democratic public culture', *Rhetoric Review* 20 (2001): 37–42

Lucas, G., 'Star Wars script (revised fourth draft)', 15 Jan 1976

Lucie-Smith, E., 'A 3D triumph for the future', *The Spectator* 277 (1996): 57–8

Lumière, A. and L. Lumière, 'L'Arrivée d'un train en gare de La Ciotat', (1895)

Lutz, W. W., 'New discoveries at Michigan universities', *The Detroit News—The Passing Show*, Sunday 23 Feb 1964, 1

Maass, U., Pat. No. US5865519 (A) 'Device for displaying moving images in the background of a stage' (1999), assigned to Musion

Markov, V. B., 'Display and applied holography in culture development', in: *Holography: Commemorating the 90th Anniversary of the Birth of Dennis Gabor* (Bellingham, WA: SPIE, 1990), pp 268–304

Markov, V. B., 'Holography in Museums – Why Not Go 3-D', *Museum* 44 (1992): 83–6

Markov, V. B. and G. I. Mironyuk, 'Holography in museums of the Ukraine', presented at *Three-Dimensional Holography: Science, Culture, Education*, Kiev, USSR, 1989

Markov Vladimir, B., 'Holography in museums ', *The Imaging Science Journal* 59 (2011): 66–74

Marx, J., 'How the "Queen Science" lost her crown: a brief social history of science fairs and the marginalization of social science', *Sociation Today* 22 (2004): http://www.ncsociology.org/sociationtoday/v22/science.htm

Massie, K. and S. D. Perry, 'Hugo Gernsback and radio magazines: an influential intersection in broadcast history', *Journal of Radio Studies* 9 (2002): 264–81

McEwen, J., 'Art brought to light', *The Spectator*, 16 Apr 1982: 25

McGrew, S., 'Mass produced holograms for the entertainment industry', *Proceedings of SPIE* 391 (1983): 19–20

McHelone, J. P., 'The signature of light: photo-sensitive materials in the nineteenth century', in: A. Thomas, M. Braun and National Gallery of Canada (eds.), *Beauty of Another Order: Photography in Science* (New Haven, CT: Yale University Press, 1997), pp 60–75

McKibbin, R., 'Work and hobbies in Britain, 1880–1950', in: J. Lerner (ed.), *The Working Class in Modern British History: Essays in Honour of Henry Pelling* (Cambridge: Cambridge University Press, 1983), pp 143–5

Mitchell, D. J., 'Reflecting nature: chemistry and comprehensibility in Gabriel Lippmann's "physical" method of photographing colours', *Notes and Records of the Royal Society* 64 (2010): 319–37

Moffat, S., 'Science: the miracle of the laser beam brings new dimensions to the world of photography', *Boy's Life*, Jun 1969: 19

Molson, F. J., 'The boy inventor in American series fiction: 1900–1930', *Journal of Popular Culture* 28 (1994): 31–48

Moorhouse, H. F., 'The work ethic and leisure activity: the hot rod in post-war America', in: P. Joyce (ed.), *The Historical Meaning of Work* (New York: Cambridge University Press, 1987), pp 257–81

Morgan, A., 'Treasures trapped in light + exhibit of Ukrainian holograms in York', *History Today* 39 (1989): 5

Morris, C. and G. Endfield, 'Exploring contemporary amateur meteorology through an historical lens', *Weather* 67 (2012): 4–8

Munir, K. A. and N. Phillips, 'The birth of the "Kodak Moment": institutional entrepreneurship and the adoption of new technologies', *Organization Studies* 26 (2005): 1665–87

Murray, R., 'Holography at the Royal College of Art, London', *International Symposium on Display Holography* 1600 (1991): 237–9

Nauman, B., *Artforum*, 9 (4), Dec 1970: cover art

Naylor, G., *Red Dwarf*, (BBC Two, 1988–99)

Newswanger, C. D., 'Dot matrix technique for generating diffraction grating patterns', presented at *Fifth International Symposium on Display Holography*, Lake Forest, IL, 1994

New York Times News Service, 'Satellite search a hobby for home: Los Alamos scientists build watching post based on simple short wave unit', *New York Times*, 20 Oct 1957, 8

Norling, J. A., 'The stereoscopic art', *Journal of the Society of Motion Picture and Television Engineers* 60 (1953): 268–308

Novotny, G. V., 'The little train that wasn't', *Electronics*, 37 (30), 30 Nov 1964: 86–9

Nuttall, G., 'Made in Britain: the new millionaires who beat the slump', *Sunday Times Magazine*, 17 Apr 1983, 31–6

Osmundsen, J. A., 'Scientists' camera has no lens', *New York Times*, 5 Dec 1963, 55

Pappu, R. and W. Plesniak, 'Haptic interaction with holographic video images', *Proceedings of SPIE* 3293 (1998): 38–45

Paramount Domestic Television, *Star Trek: The Next Generation*, (CBS, 1987–94)

Paramount Domestic Television, *Star Trek: Deep Space Nine*, (CBS, 1993–99)

Paramount Domestic Television, *Star Trek: Voyager*, (CBS, 1995–2001)

Paul Raymond Publications, '3D Hologram Issue: amazing hologram 3D centre set with free sex spex', *Men Only*, 48 (4–6), Apr–Jun 1983: cover

Pearson, J., 'The Army's new eyes: U-M Scientists develop far-sighted radar unit', *Detroit Free Press*, 20 Apr 1960, 2

Pepper, A., 'Expansion of holographic film products', *Holosphere*, 15 (1), Spring 1987: 12–3

Pepper, A., 'A single beam recording system for creative holography and education', *Proceedings of SPIE* 2043 (1994): 126–9

Pepper, A., 'When did creative holography die?', *Journal of Holography and Speckle* 3 (2006): 80–3

Phillips, N. J., H. Heyworth and T. Hare, 'On Lippmann's photography', *Journal of Photographic Science* 32 (1984): 158–69

Pihlainen, K., 'Cultural history and the entertainment age', *Cultural History* 1 (2012): 168–79

Pinch, T. J. and W. E. Bijker, 'The social construction of facts and artifacts: or how the sociology of science and the sociology of technology might benefit each other', in : W. E. Bijker, T. P. Hughes and T. J. Pinch (eds.), *The Social Construction of Technological Systems* (Cambridge, MA: MIT Press, 1987), pp 17–50

Plesniak, W. J. and M. A. Klug, 'Tangible holography: adding synthetic touch to 3D display', *Proceedings of SPIE* 3011 (1997): 53–60

Popper, F., 'The place of high-technology art in the contemporary art scene', *Leonardo* 26 (1993): 65–9

Pribram, K. H., 'The neurophysiology of remembering', *Scientific American* 220 (1969): 73–86

Rajchman, J. A., 'Promise of optical memories', *Journal of Applied Physics* 41 (1970): 1376–83

Rashed, R., 'A pioneer in anaclastics: Ibn Sahl on burning mirrors and lenses', *Isis* 81 (1990): 464–91

Reimers, F., 'Albert Vinicio Baez and the promotion of science education in the developing world 1912–2007', *Prospects* 37 (2007): 369–81

Reinert, J., 'The picture that hangs in midair', *Science Digest* 60 (1966): 30–4

Robinson, M. L. A. and A. P. Saunders, 'Making holograms at school', *School Science Review* 55 (1973): 260–73

Rodchenko, A., 'Response to Boris Kushner letter in *Sovetsko foto*, Novyi LEF no. 10, 18 Aug 1928', in: J. Gambrell (transl.) and A. N. Lavrentiev (ed.), *Aleksandr Rodchenko: Experiments for the Future* (New York: Museum of Modern Art, 2005), pp 207–12

Rolfe, L. and N. Lennon, 'Holography – new light on a new dimension', *Delta Airlines Sky*, Dec 1974: 23–9

Ross, J., 'Press release: Jonathan Ross presents LANDSCAPES & METAMORPHOSES: an exhibition of work by Jeffrey Robb,' London, 1993

Rowlands, J. J. and J. R. Killian Jr, 'Seeing solid: the third dimension at work and play', *Technology Review*, 39 (5), Mar 1937: 191–5

Rycroft, S., 'The nature of Op Art: Bridget Riley and the art of nonrepresentation', *Environment and Planning D: Society and Space* 23 (2005): 351–71

Sadowski, P., 'Holographic child's play', *Holosphere*, 15 (1), Spring 1987: 17

Sanderson, M., 'Research and the firm in British industry, 1919–39', *Social Studies of Science* 2 (1972): 107–51

Sant'Andrea, J., 'Messages of a multi-media maker', *Business Screen* 32 (1971): 13–6

Schroter, J., 'Technologies beyond the still and the moving image: the case of the multiplex hologram', *History of Photography* 35 (2011): 23–32

Schuenzel, E. C. and R. L. Moore, 'Credit card system', *IBM Technical Disclosure Bulletin* 13 (1970): 176–7

Schuster, L. F., 'Memorandum to the Director General (BBC)', British Broadcasting Corporation, 28 Apr 1942

Schuyten, P. J., 'Technology: Unmet promise of holography', *New York Times*, 24 Apr 1980, D1

Schwartz, D. and M. Griffin, 'Amateur photography: the organizational maintenance of an aesthetic code', in: T. Lindlof (ed.), *Natural Audiences: Qualitative Studies of Media Uses and Effects* (Norwood, NJ: Ablex, 1987), pp 198–224

Schwarz, D., 'Camera club photo-competitions: An ethnographic approach to the analysis of a visual event', *Research on Language and Social Interaction* 21 (1987): 251–81

Scientific American magazine, 'An author's guide to *Scientific American*,' New York, 1949

Shaikhet, A., 'Sorevnovanie foto-reporterov razvertyvaetsia [The competition of the photojournalists unfolds]', *Sovetskoe foto* (1929): 713

Sheets-Pyenson, S., 'Popular science periodicals in Paris and London: The emergence of a low scientific culture, 1820–1875', *Annals of Science* 42 (1985): 549–72

Sherman, R. A., 'Benefits to vision through stereoscopic films', *Journal of the Society of Motion Picture and Television Engineers* 61 (1953): 295–308

Shinn, T. and B. Joerges, 'The transverse science and technology culture: dynamics and roles of research-technology', *Social Science Information* 41 (2002): 207–51

Siebert, L. D., 'Large-scene front-lighted hologram of a human subject', *Proceedings of the IEEE* 56 (1968): 1242–3

Slayton, R., 'From death rays to light sabers: making laser weapons surgically precise', *Technology and Culture* 52 (2011): 45–74

Smith, C. W., '3-D or not 3-D', *New Scientist* 102 (1984): 40–4

Smith, H. and B. Van der Horst, 'Scenes', *Village Voice*, 14 Feb 1974, 2

Southall, J. P. C., 'Leonard Thompson Troland', *Journal of the Optical Society of America* 22 (1932): 509–11

Stebbins, R. A., 'Science amateurs? Rewards and costs in amateur astronomy and archaeology', *Journal of Leisure Research* 13 (1981): 289–304

Steinman, M., 'Tools to reach beyond the bounds of man's vision', *Life*, 61 (26), 23 Dec 1966: 104–19

St.-Hilaire, P., M. Lucente, J. D. Sutter, R. Pappu, C. D. Sparrell and S. A. Benton, 'Scaling up the MIT holographic video system', *Proceedings of SPIE* 2333 (1995): 374–280

Stong, C. L., 'The Amateur Scientist: mostly about how to study artificial satellites without complex equipment', *Scientific American*, 201 (1), Jan 1958: 98–109

Stong, C. L., 'The Amateur Scientist: how a persevering amateur can build a gas laser in the home', *Scientific American*, 211 (9), Sep 1964: 127–34

Stong, C. L., 'The Amateur Scientist: more about the home-made laser', *Scientific American*, 213 (12), Dec 1965: 106–13

Stong, C. L., 'The Amateur Scientist: how to compute the orbits of space vehicles', *Scientific American*, 220 (1), Jan 1969: 123–30

Stong, C. L., 'The Amateur Scientist: how to construct an argon gas laser with outputs at several wavelengths', *Scientific American*, 220 (2), Feb 1969: 118–25

Stong, C. L., 'The Amateur Scientist: how to construct a dye laser', *Scientific American*, 222 (2), Feb 1970: 116–23

Stong, C. L., 'The Amateur Scientist: how to construct an infrared diode laser', *Scientific American*, 227 (2), Mar 1973: 114–21

Stong, C. L., 'The Amateur Scientist: how to construct a nitrogen laser', *Scientific American*, 229 (6), Jun 1974: 122–7

Stroke, G. W., 'Breakthrough in "lensless photography" sets stage for 3-dimensional home color television and a possible multi-million industrial explosion in electro-optics', Press release from Univ. of Michigan for March 14–8 1966 OSA meeting,

Stroke, G. W., 'White-light reconstruction of holographic images using transmission holograms recorded with conventionally-focused images and "in-line" background', *Physics Letters* 23 (1966): 325–7

Stroke, G. W., 'Sharpening images by holography', *New Scientist* 51 (1971): 671–4

Stroke, G. W. and A. E. Labeyrie, 'White-light reconstruction of holographic images using the Lippmann-Bragg diffraction effect', *Physics Letters* 20 (1966): 368–70

Sullivan, W., 'New photo technique projects a world of 3-dimension views', *New York Times*, 19 Mar 1967, 1

Sweezey, C. O., 'All that's 3D is not holography: Disney World's Haunted House', *Holosphere*, 10 (1), Jan 1981: 1

Tanner, P. G. and T. E. Allibone, 'The patent literature of Nobel laureate Dennis Gabor (1900–1979)', *Notes and Records of the Royal Society of London* 51 (1997): 105–20

Terzian, S. G. and L. Shapiro, 'Corporate science education: Westinghouse and the value of science in mid-twentieth century America', *Public Understanding of Science* 24 (2013): 1–20

The Green Art Gallery & New England Sculptor's Guild, 'Timeless Visions: Holograms from the Collection A D 2000,' Guildford, CT, 1999

The Hologram Place, 'Press release: Lasers in Oxford Street – Holograms in the Home!,' London, 1978

The Science Talent Institute, 'Teen-age scientists offer cures for threats to U.S. technological survival', press release, 5 Mar 1956, SIA RU7091, Box 330 folder 5

Tomaszkiewicz, F., 'Continuing laser-imaging program of instruction in a public middle school', presented at *International Symposium on Display Holography*, Lake Forest, IL, 1991

Tomaszkiewicz, F., 'Holography in public middle schools: perspectives on student safety and environmental impact', presented at *Fifth International Symposium on Display Holography*, Lake Forest, IL, 1994

Tomaszkieweicz, F., 'Ten-year perspective on display holography in the middle school classroom', presented at *Sixth International Symposium on Display Holography*, Lake Forest, IL, 1997

Tucker, J., 'The historian, the picture, and the archive', *Isis* 97 (2006): 111–20

Twist, S. H., 'The picture of the future: a reverie', *New York Clipper*, 17 Aug 1912, 7

Uhrin, R., 'Crystals could unlock an old dream for data storage volume holographic storage employing photofractives promises a new era of images', *Photonics Spectra* 27 (1993): 87

Upatnieks, J. and E. N. Leith, 'Lensless photography', presented at *SPIE Photo-Optics Workshop Meeting*, Los Angeles, 1964

Upatnieks, J. and E. N. Leith, 'Lensless three-dimensional photography by wavefront reconstruction', presented at *Optical Society of America Spring Conference*, Washington DC, 1964

Upatnieks, J. and E. N. Leith, 'Lensless, three-dimensional photography by wavefront reconstruction', *Journal of the Optical Society of America* 54 (1964): 579–80
U.S. Dept. of Agriculture and State Agricultural Colleges, 'Cooperative extension work in agriculture and home economics – boys and girls radio clubs,' 1922
van der Osten, R., 'Four generations of Tom Swift: ideology in juvenile science fiction', *The Lion and the Unicorn* 28 (2004): 268–83
Veteto, B., 'Laser may make 3-D TV possible', *Dallas Times Herald*, 31 Mar 1965, A–31
von Hentig, H., 'Education and the cinematograph', *Moving Picture World* 10 (1911): 973–4
Wade, N., 'Op art and visual perception', *Perception* 7 (1978): 21–46
Walker, A. L. and R. K. Moulton, 'Photo albums: images of time and reflections of self', *Qualitative Sociology* 12 (1989): 155–82
Walker, J., 'The Amateur Scientist: how to stop worrying about vibrations and make holograms viewable in white light', *Scientific American*, 260 (5), May 1989: 134–7
Walton, S., 'Holography gets closer to your neighborhood theatre', *Popular Mechanics* 148 (1977): 94
Ward, B., 'House of tomorrow', *Radio-Electronics* (1967): 21–3
Westwood Pacific, *Command and Conquer: Red Alert 2* videogame (Microsoft, 2000)
Whittaker, J., 'On the beam . . . in color!', *Detroit Free Press*, 10 May 1970, 4–D
Wilfong, J., 'Hologram product display announced', *Conductron Antenna*, 2 Dec 1968: 4
Williams, T. R., 'Albert Ingalls and the ATM movement', *Sky and Telescope*, 81, Feb 1991: 140–2
Wilson, H., 'Conference speech, Labour Party Annual Conference', Scarborough, 1 Oct 1963
Wolff, M., 'The birth of holography: a new process creates an industry', *Innovations* (7) (1969): 4–15
Woolfgar, S., C. Coopmans and D. Neyland, 'Does STS mean business?', *Organization* 16 (2009): 5–30
Wuerker, R., 'Holography of art objects', *Proceedings of SPIE* 215 (1980): 167–71
Yaroslavsky, L. P., 'Digital holography: 30 years later', *Proceedings of SPIE* 4659 (2002): 1–11
Zellerbach, G. A., 'Selling holographic art: an analysis of past, present and future markets', *Proceedings of SPIE* 1600 (1991): 446–54
Zellerbach, G. A., 'Comments on the Market for Holographic Art', *Leonardo* 25 (1992): 419–20
Zernike, F., 'How I discovered phase contrast', *Nobel Lecture, Nobel Prize for Physics* (1953)
Zimmerman, P. R., 'Professional results with amateur ease: the formation of amateur filmmaking aesthetics 1923–1940', *Film History* 2 (1988): 267–91
Zyabrev, V. A., 'Holography: a promise for information science', *Nauchno Tekhnicheskaya Informatsiya, Seriya* 1 (1971): 30–4

UNPUBLISHED COMMUNICATIONS

Baez, A. V. to S. F. Johnston, email, 13 Mar 2003, SFJ collection
Begleiter, E. to S. F. Johnston, interview, 10 Jul 2003, Cambridge, MA, SFJ collection
Benyon, M. to S. F. Johnston, interview, 21 Jan 2003, Santa Clara, CA, SFJ collection
Benyon, M. to S. F. Johnston, email, 2 Feb 2003, SFJ collection,
Bjelkhagen, H. I. to S. F. Johnston, interview, 18–9 Sep 2002, SFJ collection
Born, M. to D. Gabor, letter, 21 Feb 1951, IC GABOR MB / 10 / 3
Casdin-Silver, H. to S. F. Johnston, interview, 3 Jul 2003, Boston, MA, SFJ collection
Charnetski, C. to S. F. Johnston, interview, 3 Sep 2003, Ann Arbor, MI, SFJ collection
Cochran, G. D. to S. F. Johnston, interview, 6 and 8 Sep 2003, Ann Arbor, MI, SFJ collection
Corey, L. E. to C. L. Stong, letter, 14 Nov 1966, ACNMAH 0012, Box 23, folder 7

Cross, L., 'The Story of Multiplex', transcription from audio recording, Spring 1976, Naeve collection
Cross, L. to S. F. Johnston, email, 25 Oct 2003, SFJ collection
Davis, W., 'Make your own telescope', promotion letter to newspaper editors, n.d., 1930, SIA RU7091, Box 120, folder 9
Denisyuk, Y. N. to S. F. Johnston, email, 3 May 2003, SFJ collection
Denton, F. to S. F. Johnston, telephone, 1 May 2003, SFJ collection
DiBiase, V. to S. F. Johnston, interview, 21 Jan 2003, Santa Clara, CA, SFJ collection
Funkhouser, A. to S. F. Johnston, email, 9 Apr 2003, SFJ collection
Gabor, D. to W. L. Bragg, letter, 6 Jun 1948, Rugby, IC GABOR EL/1
Gabor, D. to W. L. Bragg, letter, 7 July 1948, IC GABOR EL/1
Gabor, D. to T. Raison, letter, 15 Jul 1961, IC GABOR MN/5–6
Gabor, D. to G. L. Rogers, letter, 5 Aug 1966, Sci Mus M S 1014/15
Gillespie, D. and JODON Ltd to C. L. Stong, letter, 11 Nov 1966, ACNMAH 0012, Box 23, folder 7
Gillespie, D. to S. F. Johnston, interview, 29 Aug and 4–6 Sep, 2003, Ann Arbor, MI, SFJ collection
Gonzales, N. to S. F. Johnston, email, 19 Nov 2013, SFJ collection
Green, D. to S. F. Johnston, emails, 7–14 Aug 2014, London, SFJ collection
Haines, K. to S. F. Johnston, interview, 21 Jan 2003, Santa Clara, CA, SFJ collection
Hunt, S. to S. F. Johnston, email, 13 Dec 2002, SFJ collection
Ingalls, A. G. to B. W. Powell, letter, 10 Apr 1953, ACNMAH 0175, Box 8, folder 2
Jeong, T. H. to S. F. Johnston, interview, 21 Jan 2003, Santa Clara, CA, SFJ collection
Kontnik, L. T. to S. F. Johnston, interview, 20 Nov 2003, Vancouver, SFJ collection
Leith, E. N. to S. F. Johnston, interview, 22 Jan 2003, Santa Clara, CA, SFJ collection
Leith, E. N. to S. F. Johnston, interviews, 29 Aug–12 Sep 2003, Ann Arbor, MI, SFJ collection
Leith, E. N., D. Gillespie and B. Athey to S. F. Johnston, interview, 11 Sep 2003, Ann Arbor, MI, SFJ collection
Le Poole, J. B. to D. Gabor, letter, 21 Jan 1948, IC GABOR EL/1
Lohmann, A. W. to S. F. Johnston, fax, 13 May 2003, SFJ collection
McGrew, S. to J. Ross, interview, Dec 1980, Los Angeles, CA, Ross collection
McGrew, S., 'My reflections on holography', unpublished manuscript, 5 Dec 2004, SFJ collection
Moore, L. to J. Ross, interview, 1980, Los Angeles, CA, Ross collection
Munday, R. to S. F. Johnston, interview, 31 Mar 2004, Richmond, UK, SFJ collection
Nicholson, A. M. to S. F. Johnston, interview, 21 Jan 2003, Santa Clara, CA, SFJ collection
Rogers, G. L. to A. V. Baez, letter, 19 Jul 1956, Sci Mus ROGRS 6
Ross, J. to S. F. Johnston, interview, 3 Apr 2003, London, SFJ collection
Ross, J. to S. F. Johnston, telephone interview, 1 Aug 2014, London, SFJ collection
Rowe, D. F. to C. L. Stong, letter, 4 Feb 1971, ACNMAH 0012, Box 30, folder 6
Siebert, L. to S. F. Johnston, interview, 4 Sep 2003, Ann Arbor, MI, SFJ collection
Siegel, K. M., 'Speech to Conductron Missouri employees', 1966, Charnetski collection
Stong, C. L. to H. H. Larkin, letter, 11 Sep 1951, ACNMAH 0012, Box 1, folder 1
Stong, C. L. to H. Morgenroth, letter, 2 Feb 1955, ACNMAH 0012, Box 4, folder 1
Stong, C. L. to S. Heumann, letter, 12 Jul 1966, ACNMAH 0012, Box 23, folder 7
Unterseher, F. to S. F. Johnston, interview, 22 Jan 2003, Santa Clara, CA, SFJ collection
Upatnieks, J. to S. F. Johnston, email, 24 June 2003, SFJ collection
Upatnieks, J. to S. F. Johnston, email, 9 Jan 2005, SFJ collection
Zellerbach, G. A. to J. Ross, interview, 1980, Los Angeles, CA, Ross collection
Zellerbach, G. A. to S. F. Johnston, telephone interview, 12 Jun 2014, SFJ collection

INDEX

I

M

N

O

W

X

Z

The manufacturer's authorised representative in the EU for product safety is Oxford University Press España S.A. of el Parque Empresarial San Fernando de Henares, Avenida de Castilla, 2 – 28830 Madrid (www.oup.es/en or product.safety@oup.com). OUP España S.A. also acts as importer into Spain of products made by the manufacturer.

www.ingramcontent.com/pod-product-compliance
Ingram Content Group UK Ltd.
Pitfield, Milton Keynes, MK11 3LW, UK
UKHW052232270726
14060UKWH00005B/717
9780198712763